Graduate Texts in Operations Research

This series contains compact volumes on the mathematical foundations of Operations Research, in particular in the areas of continuous, discrete and stochastic optimization. Inspired by the PhD course program of the Dutch Network on the Mathematics of Operations Research (LNMB), and by similar initiatives in other territories, the volumes in this series offer an overview of mathematical methods for post-master students and researchers in Operations Research. Books in the series are based on the established theoretical foundations in the discipline, teach the needed practical techniques and provide illustrative examples and applications.

More information about this series at http://www.springer.com/series/15699

Willem K. Klein Haneveld •
Maarten H. van der Vlerk • Ward Romeijnders

Stochastic Programming

Modeling Decision Problems Under Uncertainty

Springer

Willem K. Klein Haneveld
Department of Operations
University of Groningen
Groningen, The Netherlands

Maarten H. van der Vlerk
Department of Operations
University of Groningen
Groningen, The Netherlands

Ward Romeijnders
Department of Operations
University of Groningen
Groningen, The Netherlands

Graduate Texts in Operations Research
ISBN 978-3-030-29218-8 ISBN 978-3-030-29219-5 (eBook)
https://doi.org/10.1007/978-3-030-29219-5

This Springer imprint is published by the registered company Springer Nature Switzerland AG.
The registered company address is: Gewerbestrasse 11, 6330 Cham, Switzerland

This book is dedicated to
Maarten van der Vlerk
our close collaborator and friend, who sadly
is not with us any more.

Preface

This book is based on the lecture notes that we have developed for the course Stochastic Programming for master's students in Econometrics and Operations Research at the University of Groningen. The course exists since the mid-1970s and has been running ever since, making it the longest running Stochastic Programming course worldwide. Wim Klein Haneveld was the first lecturer of the course. He made the first versions of the lecture notes. Later in the mid-1990s, they were updated and extended by Maarten van der Vlerk, who also started a similar Stochastic Programming course in Utrecht for all Ph.D. students in the Netherlands at the LNMB, the Dutch Network on the Mathematics of Operations Research. Ward Romeijnders is the current lecturer of both courses.

This book is meant to be an introduction into Stochastic Programming, particularly aimed at graduate students. The starting point of the book is a linear programming problem with random parameters, representing a decision problem under uncertainty. Several models for this problem are presented, including the main ones in Stochastic Programming: recourse models (Chap. 3) and chance constraint models (Chap. 5). We not only discuss theoretical properties of these models and algorithms for solving them but also explain the intrinsic differences between these models. At the end of the book in Chap. 8, we provide several case studies so that students can apply the theory in this book to practical problems.

Each chapter of the book is dedicated to a different model in Stochastic Programming but has a similar setup. In each chapter, we start by explaining the main concepts of the models, and we discuss small examples in full detail, to provide insights into the mathematical properties of the models. Later these properties are proved in general and presented in different theorems. We end each chapter with algorithms for solving the models.

Throughout the book, parts of the proofs of theorems are left to the reader as exercises. Other exercises in the main text are meant to support self-study using this book. More challenging theoretical assignments can be found in Chap. 7. We use these assignments as homework assignments during our course. The case studies in Chap. 8, on the other hand, are more practically oriented. They represent a wide variety of Stochastic Programming applications and are intended to be solved using a computer.

As can be expected from an introductory textbook on Stochastic Programming, large parts of it deal with two-stage recourse models and chance constraints. In addition, more emphasis than perhaps may be expected is put on stochastic mixed-integer programming (Chap. 4) and integrated chance constraints (Chap. 6), since these have been main topics in our research group in Groningen. As a result, we for example do not discuss multistage stochastic programming, distributional robustness, or abstract linear programming and the fundamental Rockafellar-Wets duality results. Therefore, we do not claim that this book covers the entire field of Stochastic Programming. This has never been the purpose of this book.

The mathematical background required to read this book is purposefully kept at a minimum. For example, we do not assume any knowledge on functional analysis or on measure theory. These subjects are not needed to understand the results in this book. We do, however, assume that readers have a basic understanding of elementary analysis and integration theory. The use of probability theory is essential, but only at an elementary level. In fact, the most general ω in this book is a finite-dimensional vector of random variables. In Appendix A we deal with some issues of probability that are used in this book. Of course, the duality theory of linear optimization is important in this book, as are the basics of convex analysis and optimization. For those who might need it, we added Appendices B and C.

With respect to notation we allow ourselves some inconsistencies throughout this book. For example, we typically omit the transposition signs that appear often in, e.g., linear programming problems. Instead we assume that not every vector is a column vector, so that the inner product cx is well defined if we interpret $c \in \mathbb{R}^n$ as a row vector and $x \in \mathbb{R}^n$ as a column vector. However, in cases where there may be confusion we do use the transposition sign. Moreover, occasionally we use still another notation to represent an inner product. In particular, in the context of convex analysis in Appendix B we use the notation $\langle \cdot, \cdot \rangle$, but not in other parts of the book.

Finally, we would like to acknowledge that the notation for the different models is to a large extent taken from the management system SLP-IOR from Kall and Mayer [13] and the corresponding SLP-IOR User's Guide. For many years, students of our Stochastic Programming course have used the SLP-IOR software to solve the case studies.

Groningen, The Netherlands

Willem K. Klein Haneveld
Ward Romeijnders

Contents

Chapter 1
Introduction

1.1 Linear Programming with Random Parameters

Linear programming has proven to be a suitable framework for the quantitative analysis of many decision problems. The reasons for its popularity are obvious: many practical problems can be modeled, at least approximately, as linear programs, and powerful software is available. Nevertheless, even if the problem has the necessary linear structure it is not sure that the linear programming approach works. One of the reasons is that the model builder must be able to provide numerical values for each of the coefficients. But in practical situations one often is not sure about the 'true' values of all coefficients. Usually the uncertainty is exorcised by taking reasonable guesses or maybe by making careful estimates. In combination with sensitivity analysis with respect to the most inaccurate coefficients this approach is satisfactory in many cases. However, if it appears that the optimal solution depends heavily on the value of some inaccurate data, it might be sensible to take the uncertainty of the coefficients into consideration in a more fundamental way. Since an evident framework for the quantitative analysis of uncertainty is provided by probability theory it seems only natural to interpret the uncertain coefficient values as realizations of random variables. This approach characterizes *stochastic linear programming*.

In this subarea of mathematical programming it is usually assumed that the joint distribution of the random coefficients is known. As we will see, a wide range of meaningful models can be formulated then. From the point of view of applications, however, this assumption is fairly demanding, since one is seldom so lucky that the available data are sufficient for a reliable estimation of the joint distribution. Nonetheless it is our opinion that stochastic programming is valuable, also in view of applications. A first reason is that some of the stochastic programming models do not need a complete knowledge of the distribution. Moreover, by means of a minimax approach sometimes the worst-case distribution compatible with the partial knowledge about the distribution can be determined (see e.g. Dupačová [5]).

W. K. Klein Haneveld et al., *Stochastic Programming*, Graduate Texts in Operations Research, https://doi.org/10.1007/978-3-030-29219-5_1

But the main reason is that in our opinion even a subjective specification of the joint distribution may serve the purpose of the decision model, i.e. the analysis of a decision problem under uncertainty, very well. One might object, of course, that subjective probabilities introduce a kind of arbitrariness in the model. We want to stress, however, that in this respect there is no fundamental distinction with applied deterministic linear programming. Also in that case the specification of the coefficients often is arbitrary to a certain extent, and it depends on the judgement of the model builder as well. In fact, there is more subjectivity in applications than sometimes is recognized. In many cases, for instance, there are several conflicting objectives, and in a linear programming framework usually all but one are represented as (goal-)restrictions, whereas the last one is used as the objective function. Hence, a choice has to be made as to the way of representing the goals in the model. In stochastic models this choice is even more intricate, since also the attitude to risk has to be modeled, and this can be done in several ways, as we will see. But again we claim that with respect to subjectivity there is only a gradual difference between deterministic and stochastic models.

Let us assume, therefore, that a linear programming problem has been completely specified, apart from some coefficients that are random variables with a joint known distribution. Traditionally, two types of models for this situation are considered in stochastic programming, namely the *wait-and-see* and the *here-and-now* models. In the first case the decision maker is assumed to be able to wait for the realization of the random coefficients, whereas in the second case the decisions have to be made before or at least without knowledge of these realizations. In the here-and-now situation a fundamental difficulty comes up: since some coefficients are undetermined the usual meaning of feasibility and/or optimality of the decisions is useless. Therefore additional specification is needed. At stake are the questions of how to measure 'risk' and how to incorporate 'risk aversion' in the decision model. In this respect, useful concepts are chance constraints, recourse actions and utility functions. For the meaning of these concepts we refer to the corresponding Chaps. 5, 3 and 2.

1.2 Examples

In this section we discuss three small examples of linear programming problems in which some coefficients are random variables. We consider problems where these random variables are in the matrix of coefficients, the right-hand side vector, and in the cost vector, respectively. For each example, we illustrate how 'risk' can be measured and incorporated in the decision problems.

1.2.1 Blending Problem

Consider a blending or mixing problem in which we have to combine two ingredients (in amounts x_1 and x_2) in a cost efficient way to obtain a sufficiently nutritious mix. In particular, the mix should contain 7 kg calcium and 4 kg protein. We assume that the corresponding linear programming problem (cf. Kall [11]) is given by

$$\begin{aligned} \min\{x_1 + x_2 : \ & \omega_1 x_1 + x_2 \geq 7 \\ & \omega_2 x_1 + x_2 \geq 4 \\ & x_1 \geq 0, x_2 \geq 0\}, \end{aligned} \tag{1.1}$$

where $x_1 + x_2$ represents the costs of the mix, and the two constraints ensure that the mix contains sufficient calcium and protein, respectively. Here, the amounts of calcium ω_1 and protein ω_2 in 1 kg of the first ingredient are independent uniformly distributed random variables with supports $[1, 4]$ and $[1/3, 1]$, respectively, whereas for simplicity it is assumed that the amounts of calcium and protein in 1 kg of the second ingredient are deterministic ($= 1$).

Interpretation as *Wait-and-See Problem (Distribution Problem)*

Suppose that it is possible to decide about the mix of ingredients x after observing the random vector $\omega = (\omega_1, \omega_2)$. Then, of course, the optimal mix of ingredients $(x_1^\star(\omega), x_2^\star(\omega))$ depends on ω, as well as the optimal value $v^\star(\omega) = x_1^\star(\omega) + x_2^\star(\omega)$.

Exercise 1.2.1 Calculate the optimal mix of ingredients $x_1^\star(\omega), x_2^\star(\omega)$, and the corresponding costs $v^\star(\omega)$ of this mix for all possible nutritious values $\omega \in \Omega = [1, 4] \times [1/3, 1]$ of ingredient 1. ◁

In principle, it is possible to calculate the distribution of $x^\star(\omega)$ and $v^\star(\omega)$, or characteristics of them, such as their mean value, variance, extreme values, etc. This is what the distribution problem is about.

Exercise 1.2.2 Calculate the density of $v^\star(\omega)$ and its expected value. ◁

Interpretation as *Here-and-Now Problem (Decision Problem)*

If one has to decide about the mix of ingredients x before knowing the actual nutritious value ω of ingredient 1, then the LP model in (1.1) needs additional specification. Below we discuss several alternatives. In all cases we assume that the distribution of ω is known.

(i) *Abolish uncertainty* by making *an appropriate guess* $\hat{\omega}$ for the nutritious value ω of ingredient 1, then solve the deterministic LP in (1.1) with ω replaced by

Table 1.1 Optimal mix of ingredients $\hat{x}$ and corresponding optimal costs $\hat{v}$ for guesses $\hat{\omega}$ of the nutritious value of ingredient 1

Guess	$\hat{\omega}$	$\hat{x}_1$	$\hat{x}_2$	$\hat{v}$
'Unbiased'	$\mathbb{E}[\omega] = (5/2, 2/3)$	18/11	32/11	50/11
'Pessimistic'	(1, 1/3)	0	7	7
'Optimistic'	(4, 1)	4	0	4

$\hat{\omega}$. Obviously, different choices of $\hat{\omega}$ lead to different decisions $\hat{x}_1$ and $\hat{x}_2$ about the mix of ingredients.

Exercise 1.2.3 Verify the decisions $\hat{x}$ for the three different guesses of $\hat{\omega}$, presented in Table 1.1. ◁

The advantage of this approach is that the reformulated problems are deterministic LPs, and thus easy to solve. Moreover, only rough information on ω is needed. The disadvantage of the approach is that it can be risky, since when $\hat{x}$ is implemented it can be infeasible. For this reason, it makes sense to use small values of $\hat{\omega}$, for example $\hat{\omega}_1 \in [1, 5/2]$ and $\hat{\omega}_2 \in [1/3, 2/3]$.

Exercise 1.2.4 Show that the 'optimistic' solution $\hat{x}_1 = 4, \hat{x}_2 = 0$ is infeasible with probability 1. ◁

(ii) *Specify 'risk' and a maximum acceptable level of 'risk'* using *chance constraints*. Below two variants of such type of constraints are presented.

- Choose (individual) reliability levels $\alpha_1, \alpha_2 \in [0, 1]$ (usually $\in (1/2, 1)$) and call $x \in \mathbb{R}^2_+$ feasible if and only if

$$\begin{cases} \Pr\{\omega_1 x_1 + x_2 \geq 7\} \geq \alpha_1, \\ \Pr\{\omega_2 x_1 + x_2 \geq 4\} \geq \alpha_2. \end{cases}$$

That is, the mix of ingredients x is called feasible if the probabilities that it contains sufficient calcium and protein, respectively, are sufficiently large. Constraints of this type are called *individual (separate)* chance constraints. Here risks are defined in terms of probability of infeasibility, i.e.,

$$\text{risk}_1 := \Pr\{\omega_1 x_1 + x_2 < 7\},$$

$$\text{risk}_2 := \Pr\{\omega_2 x_1 + x_2 < 4\}.$$

Thus, $1 - \alpha_1$ and $1 - \alpha_2$ specify the maximum acceptable risks.

Exercise 1.2.5 Show that for $x_1 > 0$ and $\alpha_1 < 1$

$$\begin{cases} \Pr\{\omega_1 x_1 + x_2 \geq 7\} \geq \alpha_1 \\ \Pr\{\omega_2 x_1 + x_2 \geq 4\} \geq \alpha_2 \end{cases} \iff \begin{cases} F_1^{-1}(1 - \alpha_1)x_1 + x_2 \geq 7 \\ F_2^{-1}(1 - \alpha_2)x_1 + x_2 \geq 4 \end{cases} \tag{1.2}$$

where F_i, $i = 1, 2$, is the cumulative distribution function (cdf) of ω_i, and

$$F_i^{-1}(\alpha) := \min_{t \in [-\infty, \infty)} \{t : F_i(t) \geq \alpha\}$$

is the α-*quantile* of (the distribution of) ω_i. ◁

The inequalities on the right in (1.2) are called the *reduced forms* of the respective individual chance constraints. They are of a type suitable for linear programming.

Remark 1.2.6 The result in Exercise 1.2.5 is based on the fact that only one random variable is contained in each constraint. Note that the result is also valid for $x_1 = 0$ (but not for $x_1 < 0$) and if $\alpha_i = 1$ (if $F_1^{-1}(0) := 1$ and $F_2^{-1}(0) := 1/3$).

- Choose a (joint) reliability level $\alpha \in [0, 1]$ (usually $\in (1/2, 1)$) and call the mix of ingredients $x \in \mathbb{R}_+^2$ feasible if and only if

$$\Pr \left\{ \begin{matrix} \omega_1 x_1 + x_2 \geq 7 \\ \omega_2 x_1 + x_2 \geq 4 \end{matrix} \right\} \geq \alpha.$$

Constraints of this type are called *joint* chance constraints. Here, risk is defined as the probability of infeasibility of the system of LP constraints, i.e.,

$$\text{risk} := \Pr \left\{ (\omega_1 x_1 + x_2 < 7) \vee (\omega_2 x_1 + x_2 < 4) \right\}.$$

In other words, the risk is defined as the probability that there is insufficient calcium, protein, or both in the mix x.
To obtain a reduced form for this joint chance constraint we observe that

$$\Pr \left\{ \begin{matrix} \omega_1 x_1 + x_2 \geq 7 \\ \omega_2 x_1 + x_2 \geq 4 \end{matrix} \right\} = \Pr \{\omega_1 x_1 + x_2 \geq 7\} \cdot \Pr \{\omega_2 x_1 + x_2 \geq 4\}$$

$$= \begin{cases} \left(1 - F_1\left(\dfrac{7 - x_2}{x_1}\right)\right) \cdot \left(1 - F_2\left(\dfrac{4 - x_2}{x_1}\right)\right), & x_1 > 0 \\ 1, & x_1 = 0,\ x_2 \geq 7 \\ 0, & x_1 = 0,\ 0 \leq x_2 < 7, \end{cases}$$

where we used that ω_1 and ω_2 are independent to obtain the first equality. We see that the reduced form of a joint chance constraint can be *nonlinear*.

At a later stage we will discuss (dis)advantages of both types of chance constraints.

(iii) *Accept infeasibility ('it is impossible to avoid risk')* but *penalize expected shortages*. For example, for the first constraint $\omega_1 x_1 + x_2 \geq 7$, the shortage of calcium in the mix is $(\omega_1 x_1 + x_2 - 7)^-$.

Remark 1.2.7 For any $z \in \mathbb{R}$, define the negative part of z by $(z)^- := \max(0, -z)$. That is, if $z \geq 0$ then $(z)^- = 0$, and if $z < 0$ then $(z)^- = -z$. If no confusion is expected, we simply write z^- instead of $(z)^-$.

Consider the following reformulation of our blending problem,

$$\min_{x \in \mathbb{R}^2_+} \left\{ x_1 + x_2 + q_1 \mathbb{E}_{\omega_1}\left[(\omega_1 x_1 + x_2 - 7)^-\right] + q_2 \mathbb{E}_{\omega_2}\left[(\omega_2 x_1 + x_2 - 4)^-\right] \right\}, \quad (1.3)$$

with unit penalty costs $q_1 > 0$ and $q_2 > 0$. In this model 'risk' is represented by expected shortages, for each constraint separately.
This is a well defined reformulation of our decision problem. However, several questions come to mind immediately: How to specify $q = (q_1, q_2)$? How to calculate the expectations involved? How to optimize over x? The following partial answers are sufficient for the moment. The calculations of the expectations are elementary but not trivial. The objective function is nonlinear but nice: it is *convex* and *(sub)differentiable*. About the specification of q we remark that it makes no sense to choose q 'small' (cf. α small in chance constrained models).

Exercise 1.2.8 Show that (1.3) can be rewritten as

$$\min_{x \in \mathbb{R}^2_+} \left\{ x_1 + x_2 + \mathbb{E}_\omega \left[\min_{y \in \mathbb{R}^2_+} \left\{ q_1 y_1 + q_2 y_2 : \begin{array}{ll} \omega_1 x_1 + x_2 + y_1 & \geq 7 \\ \omega_2 x_1 + x_2 \quad + y_2 & \geq 4 \end{array} \right\} \right] \right\}.$$

In Chap. 3 we will see that this is a so-called *simple recourse model.* ◁

1.2.2 Production Planning

Consider a production planning problem in which the variable cost of production is minimized, under the condition that the production x should satisfy the demand ω. The corresponding trivial LP is

$$\min_{x \in \mathbb{R}} \{cx : x = \omega,\ x \geq 0\}, \quad (1.4)$$

where c is the unit production cost ($c > 0$) and ω is a nonnegative continuous random variable, with $\mu = E[\omega]$, $\sigma^2 = \text{var}(\omega)$, and cdf $F(t) := \Pr\{\omega \leq t\}$, $t \in \mathbb{R}$.

Interpretation as Distribution Problem

If one is allowed to wait for the value of demand ω before the production x has to be determined, then one easily finds the optimal production level $x^\star(\omega) = \omega$ and corresponding production costs $v^\star(\omega) = cx^\star(\omega)$.

Interpretation as Decision Problem

If one has to determine the production x before the actual value of demand ω is known, then we can reformulate our linear programming problem with random variables in a similar way as for the blending problem in Sect. 1.2.1.

(i) Replace the random demand ω by the *guess* $\hat{\omega} := \mu$, or $\mu + \Delta$, where Δ is a safety stock, e.g. $\Delta = \sigma$ or $\Delta = 2\sigma$. The probability that the demand is satisfied, the so-called 'service level', is then $\Pr\{\omega \le \mu + \Delta\} = F(\mu + \Delta)$.

(ii) Obviously, a *chance constraint* $\Pr\{x = \omega\} \ge \alpha$ based on the given equality constraint is useless, since ω is continuously distributed. Moreover, for reasonable values of α_1 and α_2 the combined constraints $\Pr\{x \ge \omega\} \ge \alpha_1$, $\Pr\{x \le \omega\} \ge \alpha_2$ are also useless.

Exercise 1.2.9 Prove that for $\alpha_1 + \alpha_2 > 1$, there is no production level x satisfying both $\Pr\{x \ge \omega\} \ge \alpha_1$ and $\Pr\{x \le \omega\} \ge \alpha_2$. ◁

Hence, one has to specify priorities. This leads to the specification of a minimum required service level $\alpha \in (1/2, 1)$:

$$\min_{x \in \mathbb{R}} \{cx : \Pr\{x \ge \omega\} \ge \alpha\},$$

with solution $x^\star = F^{-1}(\alpha)$.

(iii) Another option is *penalization* of infeasibilities. Introducing unit holding costs h and shortage costs q with $-h < c < q$, we can reformulate our LP problem in (1.4) as

$$\min_{x \ge 0} \{cx + Q(x)\}, \tag{1.5}$$

where

$$Q(x) = \mathbb{E}_\omega\left[h \cdot (\omega - x)^- + q \cdot (\omega - x)^+\right]. \tag{1.6}$$

The function $Q(x)$ represents the expected costs of infeasibility, i.e., of not meeting demand ω with production x.

Remark 1.2.10 The problem (1.5) is known as the 'newsboy problem'.

Remark 1.2.11 For any $z \in \mathbb{R}$, define the positive part of z by $(z)^+ := \max(0, z)$. That is, if $z \geq 0$ then $(z)^+ = z$, and if $z < 0$ then $(z)^+ = 0$. Obviously, $(z)^+ = (-z)^-$.

Notice that $(\omega - x)^-$ and $(\omega - x)^+$ in the expression of Q in (1.6), represent the shortage and surplus in demand, respectively. However, from the production point of view, they represent production surplus and production shortage:

$$\begin{aligned} \text{demand shortage } (\omega - x)^- &= \text{production surplus } (x - \omega)^+, \\ \text{demand surplus } (\omega - x)^+ &= \text{production shortage } (x - \omega)^-. \end{aligned}$$

To solve (1.5), we use that $Q(x)$ is a convex, differentiable function, with

$$Q'(x) = -q + (q + h)F(x), \quad x \in \mathbb{R}.$$

Exercise 1.2.12 Show that the optimal solution $x^\star$, determined by $c + Q'(x^\star) = 0$, is given by

$$x^\star = F^{-1}\left(\frac{q - c}{q + h}\right).$$

◁

The optimal solution $x^\star$ may be similar to those found in the chance constraint model. Indeed, if $h = 0$, then the same solution is obtained if $q/c = 1/(1-\alpha)$. Typically, however, different specifications of risk (and corresponding costs) give rise to different optimal solutions. In fact, the difference in cost and service level of different optimal solutions may be big: whereas the prescribed service level α in the chance constraint model leads to the optimal solution $x^\star = F^{-1}(\alpha)$ (not depending on c!), the cost-optimal solution of the newsboy problem has a service level $\frac{q-c}{q+h}$. Thus, in many practical situations the cost-optimal service level is much less than often is expected (e.g., $\alpha = 0.95$).

Exercise 1.2.13 Show that (1.5) can be reformulated as

$$\min_{x \geq 0} \left\{ cx + \mathbb{E}_\omega \left[\min_{y_1, y_2 \geq 0} \{ q y_1 + h y_2 : x + y_1 - y_2 = \omega \} \right] \right\}.$$

In Chap. 3 we will see that this is a so-called *simple recourse model.* ◁

1.2.3 Portfolio Optimization

Consider an investment problem where we can invest in a 'safe' asset with known return and a 'risky' asset for which the return on investment is uncertain:

$$\max_{x \in \mathbb{R}^2_+} \Big\{ (1+r)x_1 + \omega x_2 : x_1 + x_2 \le B \Big\}. \tag{1.7}$$

Here, B denotes the available budget and $x = (x_1, x_2)$ is the portfolio determining how much we invest in the safe and risky asset, respectively. The return on investment for the safe asset is $1+r$, whereas for the risky asset it is modelled as the random variable ω with support contained in $[a, b]$, with $0 \le a < 1+r < b \le +\infty$. We let $\mu := \mathbb{E}[\omega]$ and $\sigma^2 := \mathrm{var}(\omega)$.

Interpretation as Distribution Problem

If one knows the return ω on the risky asset before the portfolio x needs to be determined, then one will either only invest in the safe or only in the risky asset, depending on whose return is largest. That is,

$$x^\star(\omega) = (x_1^\star(\omega), x_2^\star(\omega)) = \begin{cases} (B, 0), & \text{if } \omega \le 1+r, \\ (0, B), & \text{if } \omega \ge 1+r. \end{cases}$$

Moreover, the optimal return $v^\star(\omega)$ equals $v^\star(\omega) = B\max\{1+r, \omega\}$. The distribution problem is to determine the probability distribution of $x^\star(\omega)$ and $v^\star(\omega)$.

Exercise 1.2.14 Verify that $x^\star(\omega)$ has a discrete distribution with probability $p_r := F(1+r)$ attaining the value $(B, 0)$ and with probability $1 - p_r$ the value $(0, B)$. ◁

Exercise 1.2.15 Verify that $v^*(\omega)$ has a mixed distribution with $\Pr\{v^\star(\omega) = B(1+r)\} = p_r$ and $\Pr\{v^\star(\omega) \le t\} = F(t/B)$ for $t > B(1+r)$. ◁

Interpretation as Decision Problem

The essential difference with the distribution problem is that in the decision problem the portfolio x needs to be determined before the return ω on the risky asset is known. The constraints are deterministic and thus do not need further reformulation. The objective function, on the other hand, depends on ω and does need to be further specified.

It is clear that direct application of concepts as chance constraints and penalty costs, that were used in Sects. 1.2.1 and 1.2.2, does not make sense here. Nevertheless, below we use similar ideas to obtain meaningful reformulations of the stochastic objective function in (1.7).

(i) A *reasonable guess* may be to replace the return ω on the risky investment by its mean μ. Then, the optimal portfolio is $x^\star = (B, 0)$ if $\mu \leq 1 + r$ and $x^\star = (0, B)$ if $\mu \geq 1 + r$. In fact, for any guess $\bar{\omega} \in [a, b]$, we have

$$x^\star(\bar{\omega}) = \begin{cases} (B, 0), & \text{if } \bar{\omega} \leq 1 + r, \\ (0, B), & \text{if } \bar{\omega} \geq 1 + r, \end{cases}$$

similar as in the distribution problem. Interestingly, the alternative approach of maximizing the *expected* return $\mathbb{E}\left[(1 + r)x_1 + \omega x_2\right]$ is essentially the same as replacing ω by its mean μ, since the objective is linear in ω.
Thus, for individuals that are *risk-neutral* and aim to minimize the expected return only, it suffices to only know the mean μ instead of the entire distribution function F. However, for individuals that are *risk-averse* this approach is unsatisfactory, and the linear program (1.7) needs to be reformulated by also using, e.g., the variance or *expected utility* of the return. This the topic of Chap. 2.

(ii) Instead of requiring that x is feasible with sufficiently high probability, as we do in models with *chance constraints*, we will require that the portfolio x realizes a sufficiently high return with large enough reliability. We refer to this models as *P-models* and discuss two variants of them.

- Specify a minimum required return R_0 and find the portfolio x maximizing the probability that this return is realized:

$$P_1: \quad \max_{\alpha, x} \left\{\alpha : \Pr\{(1+r)x_1+\omega x_2 \geq R_0\} \geq \alpha,\ x_1+x_2 \leq B,\ x_1, x_2 \geq 0\right\}.$$

- Specify a required reliability level α_0 and maximize the return that can be obtained with at least this reliability:

$$P_2: \quad \max_{R, x} \left\{R : \Pr\{(1+r)x_1+\omega x_2 \geq R\} \geq \alpha_0,\ x_1+x_2 \leq B,\ x_1, x_2 \geq 0\right\}.$$

In both models, we are interested in the following set of feasible portfolios:

$$X_{R,\alpha} := \left\{x \in \mathbb{R}_+^2 : \Pr\{(1 + r)x_1 + \omega x_2 \geq R\} \geq \alpha,\ x_1 + x_2 \leq B\right\}.$$

Letting $\alpha^\star := \alpha^\star(R_0)$ denote the optimal solution in P_1 and $R^\star := R^\star(\alpha_0)$ the optimal solution in P_2, then all optimal solutions to P_1 and P_2 are $X_{R_0,\alpha^\star}$ and $X_{R^\star,\alpha_0}$, respectively. To characterize $X_{R,\alpha}$ we need the reduced form of $\Pr\{(1 + r)x_1 + \omega x_2 \geq R\} \geq \alpha$.

Exercise 1.2.16 Show that

$$\Pr\{(1 + r)x_1 + \omega x_2 \geq R\} \geq \alpha \quad \Leftrightarrow \quad (1 + r)x_1 + F^{-1}(1 - \alpha)x_2 \geq R.$$

◁

Using this reduced form we can characterize *all* optimal solutions to P_1 and P_2. Below, we merely specify one optimal solution for each value of R_0 and α_0. For P_1 we have

$$\begin{cases} \alpha^\star = 1,\ x^\star = (B,0), & \text{if } R_0 \leq (1+r)B, \\ \alpha^\star = 1 - F(R_0/B),\ x^\star = (0,B), & \text{if } R_0 > (1+r)B, \end{cases}$$

and for P_2 we have

$$\begin{cases} R^\star = F^{-1}(1-\alpha_0)B,\ x^\star = (0,B), & \text{if } 0 < \alpha_0 < 1 - F(1+r), \\ R^\star = B(1+r),\ x^\star = (B,0), & \text{if } 1 - F(1+r) \leq \alpha_0 \leq 1. \end{cases}$$

Notice the rapid change in optimal solution when R_0 passes the "critical" value $(1+r)B$, and α_0 the "critical" value $1-F(1+r)$. The interpretation is clear: for demanding values with $R_0 > (1+r)B$, investing in the safe asset is insufficient, and we *have to* gamble on the uncertain return of the risky asset. Moreover, in model P_1 it is then optimal to invest the entire budget B in the risky asset. Analogously, in model P_2 a return with $R > B(1+r)$ can only be achieved using the risky asset. However, the reliability in that case is only $1-F(R/B) < 1-F(1+r)$. If we want a higher reliability, then we can invest in the safe asset with fixed return.

Remark 1.2.17 From the practical point of view, one should remark that these P-models are not satisfactory. For modest values of the minimum required return R_0 or of the required reliability α_0, the optimal solution is extremely risk averse. Whereas more demanding values for these parameters lead to extremely risk seeking optimal solutions. In practice one is interested in a suitable balance between risky and safe investments, however.

Remark 1.2.18 In model P_1 there are many solutions for $aB < R_0 < (1+r)B$. This illustrates another weakness of this model. Indeed, the *probability* of the return exceeding R_0 is maximized, but we do not take into account by *how much* our return exceeds R_0. It is even possible in P_1 to select $x_1^\star + x_2^\star < B$, i.e, to not spend the entire budget. From a practical point of view, this is generally considered a suboptimal decision.

(iii) *Penalty cost model*. The idea behind this model specification is: "if the return turns out to be lower than what you need, i.e., R_0, then you have to acquire the deficit at extra costs $q > 0$ per unit". The corresponding mathematical model is

$$\max_{x \in \mathbb{R}_+^2} \left\{ \mathbb{E}_\omega \left[(1+r)x_1 + \omega x_2 - q\Big((1+r)x_1 + \omega x_2 - R_0\Big)^- \right] : x_1 + x_2 \leq B \right\}. \quad (1.8)$$

Exercise 1.2.19 Show that (1.8) can be reformulated as

$$\max_{x\in\mathbb{R}^2_+}\left\{(1+r)x_1+\mu x_2-\mathbb{E}_\omega\left[\min_{y\in\mathbb{R}_+}\{qy:\ (1+r)x_1+\omega x_2+y\geq R_0\}\right]:\right.$$
$$\left.x_1+x_2\leq B\right\}.$$

In Chap. 3 we will see that this is a so-called *simple recourse model*, for a maximization problem. ◁

Exercise 1.2.20 The model (1.8) can be solved analytically in the following way. Its objective function equals $\mathbb{E}_\omega\left[U((1+r)x_1+\omega x_2)\right]$, with $U(z) := z - q(z-R_0)^-$, $z\in\mathbb{R}$. Since U is increasing, the optimal solution $x^\star = (x_1^\star, x_2^\star)$ satisfies $x_1^\star + x_2^\star = B$, so by eliminating $x_1 := B - x_2$ and translating the random variable $\xi := \omega - (1+r)$ we get the equivalent problem

$$\max_{0\leq x_2\leq B} R(x_2), \qquad \text{where}\quad R(x_2) := \mathbb{E}_\xi\left[U\Big(B(1+r)+\xi x_2\Big)\right].$$

Use the theory of Appendices A and B to show that R is a concave function, differentiable at all $x_2\in(0,\infty)$. Find a formula for $R'(x_2)$, and an expression for the optimal value $x_2^\star$ using $R'(x_2^\star)=0$ if $x_2^\star < B$. ◁

Remark 1.2.21 In terms of the theory to be discussed in Sect. 2.1, the function U in the previous exercise can be considered as *utility function*. It is strictly increasing, and concave, so it models risk aversion, indeed. However, the utility function is rather primitive: its linearity on $(-\infty, R_0)$ and on (R_0,∞) expresses the strange phenomenon that locally on each of these intervals there is risk neutrality; the only real risk aversion is in the neighborhood of R_0.

Chapter 2
Random Objective Functions

In this chapter we discuss some classical approaches for dealing with randomness in the objective function. Since the feasible region is deterministic, we conveniently denote it by X throughout this chapter. The random objective function is denoted $g(x, \omega)$. For an example see the elementary portfolio optimization problem (1.7) in Sect. 1.2.3.

2.1 Maximization of Expected Utility[1]

Consider the following decision problem:

$$\begin{aligned}&\text{(P) Find } x \in X \text{ such that the random yield}\\ &\quad g(x, \omega) \in Z \subset \mathbb{R} \text{ is maximized, } \omega \in \Omega.\end{aligned}$$

Here, X and Ω are non-empty sets in finite-dimensional vector spaces, and the objective function g is a given real function on $X \times \Omega$. Unfortunately, the value of ω is not known at the time that the decision on x has to be taken, and therefore the problem (P) is not well defined. It is assumed that *after* the decision on x the true value of ω will be determined by a chance mechanism, and the decision maker is assumed to know the probability distribution of ω, that is, the *cdf* Φ *of* ω *is known*. Therefore, the distribution of the random yield $g(x, \omega)$ is known in principle, too. It will depend on the choice of $x \in X$. Hence, an abstract representation of (P) is

$$\text{Find } F \in \mathcal{G} \text{ which is most preferred,}$$

[1]The discussion in this section is inspired by Holloway [10], Luce and Raiffa [25], and Raiffa [34].

W. K. Klein Haneveld et al., *Stochastic Programming*, Graduate Texts in Operations Research, https://doi.org/10.1007/978-3-030-29219-5_2

where $\mathcal{G}$ is the collection of probability distributions of $g(x, \cdot)$, which arise if x takes on values in X. Of course, the preference has still to be specified, according to the original problem: maximize yields. One way to do this is to use *expected values* as criterion. In that case, $F \in \mathcal{G}$ is preferred to $F' \in \mathcal{G}$ if and only if the expected value of F is larger than that of F'. In this case (P) is specified as

$$\text{(EP) Find } x \in X \text{ such that } \mathbb{E}\left[g(x, \cdot)\right] = \int_{\Omega} g(x, \omega)\, d\Phi(\omega) \text{ is maximal.}$$

Here $\mathbb{E}$ is the expectation operator. However, the problem (EP) is not always a suitable specification of (P). For example, in this way a decision may be taken for which there is a positive probability on a high yield, but which may also lead to an extremely low yield, whereas an alternative that guarantees a reasonably large yield is not accepted. Thus, it is important to represent the decision maker's *attitude towards risk* in the specification of (P). This can be done in a systematic way by specifying a *total ordering* on

$$\mathcal{F} := \{\text{all distributions on } Z\},$$

representing the preferences of the decision maker. Even if such an ordering exists, in practice the decision maker will not be able to specify it completely. In what follows we will make a number of assumptions, which guarantee existence of the ordering mentioned. Moreover, these assumptions allow specification of a total ordering based on a limited number of questions to the decision maker.

For simplicity we assume that Z is the bounded set $[\underline{z}, \overline{z}] \subset \mathbb{R}$, and we define $A : Z \mapsto \mathcal{F}$ as follows:

$$A(z) \text{ is the degenerate distribution in } z,$$

so that the class of degenerate distributions on Z is $\mathcal{F}_0 := A(Z)$. Observe that $\mathcal{F}_0$ is the class of distributions with extremely small risk. Next we define $B : [0, 1] \mapsto \mathcal{F}$ by

$$\begin{aligned} B(p) &\text{ is the distribution that assigns probability } p \text{ to } \overline{z} \\ &\text{and the complementary probability } 1 - p \text{ to } \underline{z}. \end{aligned}$$

The class $\mathcal{F}_1 := B([0, 1])$ can be interpreted as the class of distributions with extremely large risk: $B(p)$ represents a Bernoulli trial with probability p of success (i.e., maximal yield $\overline{z}$), and probability $1 - p$ of failure (i.e., minimal yield $\underline{z}$).

We assume that the preferences of the decision maker at least satisfy the following three assumptions.

Assumption 0 (Consistency) If F_1 is preferred to F_2 and F_2 is preferred to F_3, then F_1 is preferred to F_3.

If F_1 is preferred to F_2, and the decision maker is indifferent between F_2 and F_2', then F_1 is preferred to F_2'.

If F_1 is preferred to F_2 and the decision maker is indifferent between F_1 and F_1', then F_1' is preferred to F_2.

Assumption 1 (Preference on $\mathcal{F}_0$) Let $z_1, z_2 \in Z$. Then $A(z_1)$ is preferred to $A(z_2)$ if $z_1 > z_2$.

Assumption 2 (Preference on $\mathcal{F}_1$) Let $p_1, p_2 \in [0, 1]$. Then $B(p_1)$ is preferred to $B(p_2)$ if $p_1 > p_2$.

Assumptions 1 and 2 appear to be necessary for a correct representation of the idea that large yields are preferred to smaller ones. The following assumption is crucial:

Assumption 3 (Existence) For every $F \in \mathcal{F}$ there exists an $F_0 \in \mathcal{F}_0$ that is perceived as equivalent to F by the decision maker (Notation: $F \simeq F_0$).

The meaning of $F \simeq F_0 = A(z)$ is that the decision maker is indifferent between a random yield according to the distribution F and the fixed yield z. Consistency of the preference pattern (Assumption 0) implies that F is preferred to any amount less than z (Assumption 1), but that F is less attractive than any amount strictly larger than z (Assumption 1). The amount z is called the *certainty equivalent* for F. Actually, Assumptions 0, 1, and 3 imply existence of the function $CE : \mathcal{F} \mapsto Z$ with

$$CE(F) = z \quad \Leftrightarrow \quad F \simeq A(z). \tag{2.1}$$

In particular, it holds $CE(A(z)) = z$, $z \in Z$. Assumptions 0, 1, and 3 imply that F is preferred to F' if and only if $CE(F) > CE(F')$. Now we are ready to give the complete specification of (P) in the *certainty equivalent decision problem*

$$\text{(CEP) Find } F \in \mathcal{G} \text{ that maximizes } CE(F).$$

This formulation would be of little practical value if the function CE could not be determined in such a way that it represents the risk attitude of the decision maker. However, it appears that CE is completely specified, *and can be computed*, if this function is known on the class $\mathcal{F}_1$ (the extremely risky distributions). To see this, define

$$V : [0, 1] \mapsto Z, \;\; V(p) := CE(B(p)).$$

Because $B(0) = A(\underline{z})$, it holds $V(0) = \underline{z}$. Analogously, we obtain $V(1) = \overline{z}$. Assumption 2 (which was not used before) implies that V is a strictly increasing function. This means that the inverse function $U := V^{-1} : Z \mapsto [0, 1]$ is defined, via $U(\underline{z}) = 0$, $U(\overline{z}) = 1$, and for $\underline{z} < z < \overline{z}$ by

$$U(z) = p_0 \Leftrightarrow \begin{cases} V(p) < z, \text{ if } p < p_0 \\ V(p) > z, \text{ if } p > p_0 \end{cases}$$

If we conveniently assume

Assumption 4 (Continuity) The function V is continuous.

Then it immediately follows that $U(z) = p$ if and only if $V(p) = z$. This relation between z and p means precisely that $A(z) \simeq B(p)$, that is

$$U(z) = p \Leftrightarrow V(p) = z \tag{2.2}$$

$\Leftrightarrow$ The decision maker is indifferent between the fixed yield z and the yield of the Bernoulli trial with probability of success p.

The surprising fact is now that the function U (that is, the function V) completely determines the mapping CE:

$$CE(F) = U^{-1}\left(\int U(z)\,dF(z)\right) \quad \Leftrightarrow \quad U\left(CE(F)\right) = \int U(z)\,dF(z). \tag{2.3}$$

The reasoning behind this result is straightforward (but needs more explanation than given here): if the *fixed* yield z is worth the same as the Bernoulli trial with probability of success $U(z)$, then the *stochastic* yield with distribution F must be considered to be equivalent to the Bernoulli trial with probability of success $\int U(z)\,dF(z)$.

Since U is strictly increasing, we can write the problem (CEP) somewhat more explicitly as

(P_0) Find $F \in \mathcal{G}$ such that $\int U(z)\,dF(z)$ is maximized.

On naming the function U *utility function* (following Von Neumann and Morgenstern [51]), this specification of problem (P) means that the decision maker *maximizes* his *expected utility*. Adding the utility function to the data of (P) completes the specification of the decision problem, at least if we agree that expected utility is to be maximized.

2.1.1 Specifying Utility Functions

It remains to specify the utility function. This utility function U is derived from and can be interpreted as a representation of the *decision maker's* attitude towards risk *in problem* (P). It is therefore not a function that can be derived once and for all. The assumptions imply that the function U is continuous and strictly increasing on Z, with $U(\underline{z}) = 0$ and $U(\overline{z}) = 1$. Conversely, it is easy to verify that for every function U satisfying these conditions, relation (2.3) specifies a total preference ordering in $\mathcal{F}$ (namely, F_1 is preferred to F_2 if and only if $CE(F_1) > CE(F_2)$) that satisfies Assumptions 0–4. Moreover, the utility function corresponding to CE is precisely U. Within the framework of our assumptions any function $U : Z \mapsto [0, 1]$, which is strictly increasing with $U(\underline{z}) = 0$ and $U(\overline{z}) = 1$, can serve as utility function.

Example 2.1.1 Let

$$U(z) = \frac{z - \underline{z}}{\overline{z} - \underline{z}}.$$

In this case the problem (CEP) is equivalent to (EP). However, the problem (CEP) allows specification of more suitable utility functions. ◇

Empirically Specifying Utility Functions

In principle it is possible to determine U *empirically* via V: let the decision maker specify, for sufficiently many values of p, which fixed yield $V(p)$ he or she perceives as equivalent to a Bernoulli trial with probability p on yield $\overline{z}$ and probability $1 - p$ on $\underline{z}$. (By Assumption 3 the decision maker is capable to provide such statements.) Alternatively, if one doubts the decision maker's understanding of probability calculus, the following approach can be used. Start by asking which fixed yield he or she finds equivalent to a lottery with 50% chance on $\underline{z} = V(0)$ and 50% chance on $\overline{z} = V(1)$; this will give the value of $V(1/2)$. In successive questions of this type the decision maker should come up with the certainty equivalent value of a lottery with 50% chance on $z_1 = V(p_1)$ and 50% chance on $z_2 = V(p_2)$, where p_1 and p_2 are such that the corresponding z_1 and z_2 are already known, leading to the value $V(\frac{p_1+p_2}{2})$.

Measures of Risk Aversion

An alternative to the empirical approach is to postulate a utility function based on general characteristics, for example on quantities such as *risk aversion*. A measure for the decision maker's aversion against (the risk in) the distribution $F \in \mathcal{F}$ is the *risk premium*

$$RA(F) := \mu_F - CE(F) \qquad \text{with } \mu_F := \int z\, dF(z). \tag{2.4}$$

The aversion towards F is indeed larger as $CE(F)$ is smaller. The definition is chosen such that $RA(F) = 0$ if $CE(F) = \mu_F$, that is, if the decision maker feels that the certainty equivalent of F precisely equals its expected value. A decision maker who makes this choice for all distributions F is called *risk neutral*; in this case the risk function is linear. If $RA(F) \geq 0$ for all F, then the decision maker is *risk averse*. Apparently the decision maker is willing to pay a premium $RA(F)$ to obtain a yield μ_F with certainty instead of the stochastic yield with cdf F. If $RA(F) \leq 0$ for all F, then the decision maker is called *risk seeking*. These characterizations can also be described in terms of properties of the utility function. To this end, use (2.3) to obtain an equivalent form of (2.4):

$$U\left(\mu_F - RA(F)\right) = \int U(z)\,dF(z). \tag{2.5}$$

Thus, a risk averse attitude is equivalent to

$$U\left(\int z\,dF(z)\right) \geq \int U(z)\,dF(z) \qquad \forall F \in \mathcal{F}.$$

This inequality holds for all $F \in \mathcal{F}$ if U is concave (due to Jensen's inequality, see Appendix B). Conversely, if U is not concave, then it is easy to find an $F \in \mathcal{F}$ for which the inequality does not hold. Hence, we may conclude that the decision maker is *risk averse precisely if his utility function in* (P_0) *is concave*. Analogously it follows that the decision maker is *risk seeking precisely if his utility function is convex*.

A possible *local* measure for the *risk aversion* in terms of the utility function U is

$$lra(z) := -\frac{U''(z)}{U'(z)}, \qquad z \in Z, \tag{2.6}$$

where it is assumed that U'' exists and that $U' > 0$ (see Arrow [1] and Pratt et al. [32]). It follows indeed that $lra(z) \geq 0$ if U is locally concave, and $lra(z) \leq 0$ if U is locally convex in z. An intuitive justification of this concept is based on the Taylor polynomial of U around μ_F:

$$U(z) \approx U(\mu_F) + U'(\mu_F) \cdot (z - \mu_F) + \frac{1}{2}U''(\mu_F) \cdot (z - \mu_F)^2.$$

On taking expectations with respect to F this gives

$$\int U(z)\,dF(z) \approx U(\mu_F) + \frac{1}{2}U''(\mu_F) \cdot \sigma_F^2,$$

where σ_F^2 denotes the variance of F. Together with

$$U\left(\mu_F - RA(F)\right) \approx U(\mu_F) - U''(\mu_F) \cdot RA(F)$$

Eq. (2.5) gives $RA(F) \simeq \frac{1}{2}\sigma_F^2 \cdot lra(\mu_F)$.

If one has confidence in the interpretation of *lra* and has some insight in the risk attitude of the decision maker, one can directly postulate the *lra* function for the decision maker. After solving (2.6), with side conditions $U(\underline{z}) = 0$ and $U(\overline{z}) = 1$, the utility function U then follows.

Example 2.1.2 If $lra(z) := \lambda \neq 0$ for all $z \in Z$, then $U(z) = \alpha - \beta e^{-\lambda z}$ for suitable constants α and β.

'Positive but decreasing risk aversion as yields increase' can be modeled by $lra(z) := (az + b)^{-1}$ for certain $a > 0$ and $b > -a\underline{z}$. For $a = 0$ we obtain the exponential utility function as above; if $a = 1$, then $U(z) = \gamma + \delta \log(z + b)$; otherwise $U(z) = \varepsilon + \eta(az + b)^{1-1/a}$. ◇

Remark 2.1.3 For the ease of exposition, we restricted the set of possible values of the yield to a *bounded set* $Z = [\underline{z}, \overline{z}]$ in this section. But in practice it is reasonable to take $Z = \mathbb{R}$. Similarly, in practice a utility function is a real function, mapping the yields in $Z = \mathbb{R}$ to the utilities in $\mathbb{R}$ (rather than in [0, 1]).

2.2 The Mean-Variance Model[2]

In practice the distribution F (of e.g. the random yield $g(x, \omega)$) is often characterized by its mean value $\mu(x)$ and variance $\sigma^2(x)$. If one has to choose between several distributions (e.g. for every $x \in X$ a different distribution) the choice is then exclusively determined by considering $\left\{\left(\mu(x), \sigma^2(x)\right),\ x \in X\right\}$ compared to each other. In fact, there are two goals:

(1) maximization of the average yield $\mu(x)$
(2) minimization of the variance $\sigma^2(x)$, which is taken as a measure of the risk of the corresponding distribution.

Often both goals are combined in the following way:

$$(\mathrm{P}_1)\ \max_{x \in X} \mu(x) - k \cdot \sigma^2(x),$$

where the decision maker is asked to provide the value of the parameter k (usually $k \geq 0$). The value of k can then be interpreted as a measure of the risk aversion of the decision maker. A practical advantage of this model is, that only the mean μ and the

[2]The approach outlined in this section is based on Markowitz [26].

variance σ^2 of the distribution F need to be specified; in general, this is much easier than providing a complete specification of the cdf F. The obvious consequence is that distributions which have the same μ and σ^2, but which otherwise may be very different, are considered to be equivalent.

Suppose that the yield $g(x, \omega)$ is linear in x, that is,

$$g(x, \omega) := \omega' x, \quad x \in X \subset \mathbb{R}^n,$$

where ω is a random n-vector with a known distribution that does not depend on x. We will denote the vector of mean values as $m := \mathbb{E}_\omega [\omega]$, and the covariance matrix as $V := \mathbb{E}_\omega \left[(\omega - m)(\omega - m)'\right]$. It is not surprising that the expected value $\mu(x)$ and the variance $\sigma^2(x)$ of $g(x, \omega)$ depend only on (x and) m and V, and not on other characteristics of the distribution of ω. It holds

$$\begin{aligned}\mu(x) &:= \mathbb{E}_\omega [g(x, \omega)] = m'x \\ \sigma^2(x) &:= \mathrm{var}(g(x, \omega)) = \mathbb{E}_\omega \left[(\omega'x - m'x)^2\right] \\ &= \mathbb{E}_\omega \left[x'(\omega - m)(\omega - m)'x\right] = x'Vx.\end{aligned}$$

Hence, in case of a linear yield function the problem (P_1) is

$$(P_{1L}) \max_{x \in X} m'x - k \cdot x'Vx.$$

Such a model is common for example in classical portfolio selection theory [26]. In a typical case we have $X = \{x \in \mathbb{R}^n : x \geq 0, \sum x_i = 1\}$, with x_i denoting the fraction of the budget allocated to stock i. The yield of a unit investment in i is ω_i, which is essentially unknown, except for (some characteristics of) its distribution. The problem (P_{1L}) is then a proper quadratic programming problem: it has linear constraints and a quadratic concave objective function which is to be maximized.

2.2.1 *Relation with Expected Utility Theory*

How can the model (P_{1L}) be related to the very general theory of Sect. 2.1? This can be done in two ways.

(1) Suppose the *utility function is quadratic*, that is

$$U(z) = 2\gamma z - \gamma^2 z^2, \quad 0 \leq z \leq \gamma^{-1},$$

for a given 'small' positive γ.

Exercise 2.2.1 Why $\gamma > 0$? Why $z \leq \gamma^{-1}$? ◁

With this specification the expected utility of $\omega' x$ equals

$$\mathbb{E}_\omega \left[U(\omega'x)\right] = U(m'x) - \gamma^2 \cdot x'Vx.$$

(Verify this.) We see that maximization of $\mathbb{E}_\omega\left[U(\omega' x)\right]$ leads to mean-variance efficient solutions, and hence to a solution of (P_{1L}). However, the corresponding value of the parameter k can not be determined a priori. For the quadratic utility function, the local measure for risk aversion, as defined in Sect. 2.1, is

$$lra(z) = \frac{\gamma}{1-\gamma z}, \quad 0 \le z \le \gamma^{-1}.$$

This illustrates why the quadratic utility function is not generally liked, despite its nice analytical properties:

(a) The quadratic utility function is adequate only for values of z smaller than γ^{-1}.
(b) Risk aversion as measured by $lra(z)$ is small for $z = 0$ and increases with z, to become large if $z \approx \gamma^{-1}$. This contradicts the common intuition, namely: decision makers tend to avoid high risk if the yields are low, whereas they are less averse to the same variation if higher yields (on average) are at stake.

(2) (See Freund [8]) Suppose the *utility function is exponential*:

$$U(z) = 1 - e^{-\lambda z}, \quad -\infty < z < \infty$$

for given $\lambda > 0$. For this utility function it holds $lra(z) = \lambda$, for every z. Moreover, suppose that ω is *normally distributed*, then also $g(x,\omega) = \omega' x$ is a normal random variable. The expected utility of $\omega' x$ is now

$$\mathbb{E}_\omega\left[U(\omega' x)\right] = \int_{-\infty}^{\infty} \left(1 - e^{-\lambda z}\right) \varphi(z)\, dz,$$

where φ is the pdf of the normal distribution with mean $\mu = \mu(x)$ and variance $\sigma^2 = \sigma^2(x)$, so that

$$\begin{aligned}
\mathbb{E}_\omega\left[U(\omega' x)\right] &= 1 - \int_{-\infty}^{\infty} e^{-\lambda z} \frac{1}{\sigma\sqrt{2\pi}} e^{-\frac{1}{2}\left(\frac{z-\mu}{\sigma}\right)^2} dz \\
&= 1 - \frac{1}{\sqrt{2\pi}} \int_{-\infty}^{\infty} e^{-\lambda(\mu+\sigma t)-\frac{1}{2}t^2} dt \qquad \left(\text{subst. } \frac{z-\mu}{\sigma} \equiv t\right) \\
&= 1 - e^{-\lambda\mu+\frac{1}{2}\lambda^2\sigma^2} \cdot \frac{1}{\sqrt{2\pi}} \int_{-\infty}^{\infty} e^{-\frac{1}{2}(t+\lambda\sigma)} dt \\
&= 1 - e^{-\lambda\left(\mu-\frac{1}{2}\lambda\sigma^2\right)} \\
&= U\left(\mu - \frac{\lambda}{2}\sigma^2\right).
\end{aligned}$$

We see that expected utility is maximized if $\mu - \frac{\lambda}{2}\sigma^2$ is maximal, so that problem (P_0) as currently specified is equivalent to

$$\max_{x \in X} m'x - \frac{\lambda}{2} x'Vx,$$

hence equivalent to (P_{1L}), if we identify k with $\lambda/2$.

Objections to the Mean-Variance Model

Despite its relative simplicity, there are several objections to the model (P_{1L}). We mention only two.

(i) The specification of the parameter k is *scale dependent*. It is desirable to have a dimension free parameter instead. This can be obtained e.g. by using the standard deviation (instead of the variance) as a characterization of the risk associated with a given distribution:

$$(P'_1) \max_{x \in X} \mu(x) - k \cdot \sigma(x).$$

(ii) The second objection to model (P_{1L}) concerns the measure of risk. Both the variance $\sigma^2(x)$ and the standard deviation $\sigma(x)$ are of limited use because of their inherent *symmetry*: positive deviations have the same weight as negative deviations. In many applications this is not realistic: too large yields bring about completely different consequences as too small yields. Models that allow asymmetric risk measures are e.g. the recourse models used in stochastic programming (see Chap. 3), or models based on semi-variances. (Also the P-models of stochastic programming, see the elementary portfolio optimization in Sect. 1.2.3 for example, deal with asymmetric risk measures. Nevertheless, the modeling of risk aversion in such P-models is not convincing.)

Chapter 3
Recourse Models

It is fair to say that recourse models are the most important class of models in stochastic programming, both in theory and in applications. Recourse models are reformulations of decision problems that model stochastic infeasibilities by means of corrections afterwards. First we treat this concept in a deterministic setting. The notation follows that of the model management system SLP-IOR, see Kall and Mayer [12] and the SLP-IOR User's Guide.

3.1 Models with Penalty Costs in Deterministic LP

In this setting the approach can be seen as relaxation of *soft constraints (goal constraints)*. Consider the following underlying LP problem:

$$\min_{x \in X}\{cx : Ax = b,\ Tx \sim h\}, \tag{3.1}$$

where the objective function cx and the left-hand sides of the constraints $Ax = b$ and $Tx \sim h$ are linear functions of the vector of decision variables $x \in \mathbb{R}^n$, and where c is a row vector in $\mathbb{R}^n$ representing the unit cost of x. The right-hand sides $b \in \mathbb{R}^{m_1}$ and $h \in \mathbb{R}^m$ are column vectors, and A and T are real-valued matrices of appropriate dimension. Moreover, the relational symbol $\sim$ denotes $=, \geq$, or $\leq$ (componentwisely).

The constraints $Ax = b$ are m_1 'hard' constraints (e.g. technical constraints), which we will write as equalities (inequalities can be taken care of by including suitable slack variables). The constraints denoted $Tx \sim h$ are m goal constraints, that is, we would like to satisfy them *but not at any price*. Finally, the set X specifies simple bounds on x, i.e., $X = \{x \in \mathbb{R}^n : x^l \leq x \leq x^u\}$; default is $X = \{x \in \mathbb{R}^n : 0 \leq x < \infty\} = \mathbb{R}^n_+$.

W. K. Klein Haneveld et al., *Stochastic Programming*, Graduate Texts in Operations Research, https://doi.org/10.1007/978-3-030-29219-5_3

In many modeling situations, there are several goals that we would like to achieve at the same time. In the LP model (3.1) they are represented by the objective function and m goal constraints. Which of these goals ends up in the objective, and which become goal constraints is a modeling issue.

Often the following *penalty model* is preferable over the LP model (3.1):

$$\min_{x\in X}\{cx + v(h - Tx) : Ax = b\} = \min_{x\in X}\{cx + v(z) : Ax = b,\ Tx + z = h\}, \quad (3.2)$$

where z is the vector of deviations of the goals, with components $z_i = h_i - t_i x$, where t_i is the i-th row of T, $i = 1, \ldots, m$, and $v(z)$ gives the penalty costs for deviations z. The *penalty function* indicates how the goals $0 \sim z_i, i = 1, \ldots, m$, are weighted against the objective function cx.

There are many ways to specify the penalty function v, so that it is a flexible modeling tool. In the rest of this section we discuss several possibilities.

First we note that to the LP model (3.1) corresponds an extreme specification of the penalty function:

$$v(z) = \begin{cases} 0, & 0 \sim z_i,\ i = 1, \ldots, m \\ \infty, & \text{otherwise.} \end{cases}$$

3.1.1 *Individual Penalty Costs*

An important special case is characterized by *separable (individual), linear* penalty costs, which can be different for surpluses and shortages. In this case we have

$$v(z) = \sum_{i=1}^{m} v_i(z_i), \quad z \in \mathbb{R}^m, \quad (3.3)$$

with

$$v_i(z_i) = q_i^+(z_i)^+ + q_i^-(z_i)^-, \quad z_i \in \mathbb{R},$$

where z_i is the deviation in the ith constraint.

The specification of the penalty cost parameters q_i^- and q_i^+ depends on the nature of the goal constraints. In general it holds that

- If the goal is $t_i x = h_i$, i.e. $0 = z_i$, then $q_i^- > 0$ and $q_i^+ > 0$.
- If the goal is $t_i x \geq h_i$, i.e. $0 \geq z_i$, then $q_i^- \leq 0$, $q_i^+ > 0$, with $q_i^- + q_i^+ > 0$. Thus, in this case there is a small reward if $z_i < 0$ and a large penalty if $z_i > 0$.
- If the goal is $t_i x \leq h_i$, i.e. $0 \leq z_i$, then $q_i^- > 0$, $q_i^+ \leq 0$, with $q_i^- + q_i^+ > 0$.

Exercise 3.1.1 Show that the penalty function v in (3.3) is convex if and only if $q_i^- + q_i^+ \geq 0$, $i = 1, \ldots, m$. ◁

3.1.2 Refined Individual Penalty Costs

The following *refined individual* penalty function is worked out for the case that the goal is $t_i x = h_i$, i.e. $0 = z_i$. Similar as for individual penalty costs in (3.3), we have separate penalty costs for each goal constraint, but now large deviations (outside some interval $[-l_i, u_i]$ containing 0) are penalized (much) harder than small ones.

$$v_i(z_i) = q_i^{1+}(z_i)^+ + \left(q_i^{2+} - q_i^{1+}\right)(z_i - u_i)^+ \qquad (3.4)$$
$$+ q_i^{1-}(z_i)^- + \left(q_i^{2-} - q_i^{1-}\right)(z_i + l_i)^-, \quad z_i \in \mathbb{R},$$

with $-q_i^{2-} < -q_i^{1-} < 0 < q_i^{1+} < q_i^{2+}$.

Remark 3.1.2 This penalty function can also be used to model the goal $z_i \in [-l_i, u_i]$ by taking $q_i^{1-} = q_i^{1+} = 0$.

3.1.3 Joint Penalty Costs

The following *joint* penalty function can be used if the goal is to have $0 \leq z_i$, $i = 1, \ldots, m$, and if the maximum deviation is more important than the weighted sum of the deviations.

$$v(z) = q_0 \max_{i=1,\ldots,m} (z_i)^-, \quad z \in \mathbb{R}^m, \qquad (3.5)$$

for some $q_0 \in \mathbb{R}$, $q_0 > 0$.

Exercise 3.1.3 Show that the penalty function v in (3.5) is convex if $q_0 > 0$. ◁

3.1.4 Penalty Costs Based on Recourse Actions

Next we consider a class of penalty functions, that is based on an extension of model (3.1). Later on, it will appear this class contains all important specifications, in fact. The extension works as follows. It is assumed that additional actions exist, that were neglected up to now in model (3.1), by which possible deviations in the goal constraints can be eliminated. These additional actions are called *recourse actions*, and the decision variables $y \in \mathbb{R}^m$ representing them are called *recourse variables*. Adding the recourse into model (3.1) leads to

$$\begin{array}{llllll} \min\limits_{x,y}\{ & cx & + & qy & : & \\ & Ax & & & = b & \\ & Tx & + & Wy & \sim h & \\ & x \in X & & \underbrace{y \in Y}_{(*)} & & \}, \end{array} \tag{3.6}$$

where the 'block' marked with $(*)$ is the additional recourse part. Here, W is the *recourse technology matrix* of size $m \times n$, q is the $1 \times p$ vector of *unit recourse costs*, and the set Y specifies simple bounds on y, i.e., $Y = \{y \in \mathbb{R}^p : y^l \leq y \leq y^u\}$; default is $Y = \mathbb{R}^p_+$. Together the triple (Y, q, W) specifies the *recourse structure* in the extended model.

Exercise 3.1.4 Show that (3.6) is a relaxation of (3.1), at least if $0 \in Y$. ◁

Remark 3.1.5 In (3.6) it is allowed to have additional constraints in y only. Some rows in $Tx + Wy \sim h$ may have all coefficients in T and h equal to zero; alternatively, such constraints can be part of the definition of the set Y.

Now we are ready to describe formally the penalty function v induced by the recourse structure (Y, q, W):

$$v(z) = \min_{y \in Y}\{qy : Wy \sim z\}, \quad z \in \mathbb{R}^m. \tag{3.7}$$

This penalty function is called the *recourse penalty cost function*. For $z \in \mathbb{R}^m$, the recourse function v gives, for the underlying recourse structure, the minimum recourse costs to compensate for deviations in the goal constraints $Tx \sim h$.

It is easy to see that the extended model (3.6) is equivalent to the penalty model in (3.2) if the penalty function v is specified as the recourse function in (3.7). Moreover, all previous examples of penalty functions, i.e., the individual, refined individual, and joint penalty function in (3.3)–(3.5), respectively, are a special case of (3.7). In the following Exercises 3.1.6–3.1.8 the corresponding specifications of the recourse structure (Y, q, W) are given.

Exercise 3.1.6 Consider the penalty cost function v defined in (3.7) with equality constraints and with *simple recourse* structure defined by $p = 2m$,

$$Y = \left\{(y^+, y^-) \in \mathbb{R}^p : y^+ \geq 0,\ y^- \geq 0\right\},$$

and

$$\begin{pmatrix} q \\ W \end{pmatrix} = \begin{pmatrix} q^+ & q^- \\ I & -I \end{pmatrix},$$

where I is the $m \times m$ identity matrix. Then, the penalty function v in (3.7) can be written as

$$v(z) = \min_{y^+,y^-} \{q^+ y^+ + q^- y^- : y^+ - y^- = z\}, \quad z \in \mathbb{R}^m.$$

Show that v is equivalent to the individual penalty function in (3.3) if $q^+ + q^- \geq 0$. ◁

Exercise 3.1.7 Consider the penalty cost function v defined in (3.7) with equality constraints and with *multiple simple recourse* structure defined by $p = 4m$,

$$Y = \left\{(y^{2+}, y^{1+}, y^{1-}, y^{2-}) \in \mathbb{R}^p_+ : y^{1+} \leq u,\ y^{1-} \leq l\right\},$$

and

$$\begin{pmatrix} q \\ W \end{pmatrix} = \begin{pmatrix} q^{2+} & q^{1+} & q^{1-} & q^{2-} \\ I & I & -I & -I \end{pmatrix}.$$

Show that v corresponds to the refined individual penalty function in (3.4) if $-q_i^{2-} < -q_i^{1-} < 0 < q_i^{1+} < q_i^{2+}$. ◁

Exercise 3.1.8 Consider the penalty cost function v defined in (3.7) with equality constraints and with *almost simple recourse* structure defined by $p = m + 1$, $Y = \mathbb{R}^p_+$, and

$$\begin{pmatrix} q \\ W \end{pmatrix} = \begin{pmatrix} q_0 & 0 \\ -e & I \end{pmatrix},$$

with $e = (1, \ldots, 1)' \in \mathbb{R}^m$. Show that v corresponds to the joint penalty function v defined in (3.5) if $q_0 > 0$. ◁

3.2 Recourse Models in Stochastic Linear Programming

3.2.1 Representations of Recourse Models

Suppose that the following linear programming model with random parameters in the constraints has been formulated:

$$(\mathrm{LP}_0(\omega)) \qquad \min_{x \in \mathbb{R}^n} \left\{ cx : \begin{array}{c} Ax = b \\ T(\omega)x \sim h(\omega) \\ x \in X \end{array} \right\}.$$

The set X describes lower and upper bounds on the decision variables:

$$X = \{x \in \mathbb{R}^n : x^l \leq x \leq x^u\}.$$

Usually, $x^l = 0$ and $x^u = \infty$, that is, $X \in \mathbb{R}^n_+$. Here $Ax = b$ represents m_1 deterministic equality constraints, and $T(\omega)x \sim h(\omega)$ represents m random (in)equality constraints. $T(\omega)$ is an $m \times n$ matrix, and $h(\omega)$ is an $m \times 1$ vector, both depending on a random vector $\omega \in \mathbb{R}^r$.

We assume that we are dealing with a decision problem. That is, one has to decide upon x before knowing the actual value of ω; only its distribution on the support $\Omega \subset \mathbb{R}^r$ is supposed to be known. Consequently, $\mathrm{LP}_0(\omega)$ needs reformulation of the random constraints. In recourse models, the random constraints are modelled as soft constraints: possible violation is accepted, reluctantly, but the cost of violations will influence the choice of x. In fact, a *second-stage* linear programming model is introduced, with second-stage variables $y \in \mathbb{R}^p$ that describe how violated random constraints are dealt with. The second stage is characterized by the fact, that its decisions are made *after* observation of the value of ω. This explains the name recourse model.

Standard Formulation of Recourse Models

Formally, a *recourse structure* is specified by a triple (Y, q, W) where

- $Y = \{y \in \mathbb{R}^p : y^l \leq y \leq y^u\}$, usually $Y = \{y \in \mathbb{R}^p : y \geq 0\}$, describes the feasible set of recourse actions y
- q is a $1 \times p$ vector of *unit recourse costs*
- W is an $m \times p$ matrix, the *recourse* (technology) *matrix*

Applying this recourse structure to the problem $\mathrm{LP}_0(\omega)$ leads to the following well-defined decision problem:

$$\text{(SLPwR)} \qquad \min_{x \in X} \left\{ cx + \mathbb{E}_\omega \Big[\underbrace{\min_{y \in Y} \{qy : Wy \sim h(\omega) - T(\omega)x\}}_{\text{second-stage LP}} \Big] : Ax = b \right\}$$

The relation $\sim$ in the second stage is the same relation as in the random constraints of $\mathrm{LP}_0(\omega)$.

Exercise 3.2.1 Show that the penalty cost reformulations of the blending, production planning, and portfolio optimization problem in Chap. 1 can be considered as recourse models. ◁

Compact Formulation of Recourse Models

A more compact formulation uses additional functions

$$v(z) := \min_{y \in Y}\{qy : Wy \sim z\}, \quad z \in \mathbb{R}^m, \tag{3.8}$$

$$Q(x) := \mathbb{E}_\omega \left[v(h(\omega) - T(\omega)x)\right], \quad x \in \mathbb{R}^n. \tag{3.9}$$

The function v is the *recourse penalty* cost *function* or *(second-stage) value function*; for any vector $z \in \mathbb{R}^m$ of 'deviations in the random constraints $T(\omega)x \sim h(\omega)$' it describes the corresponding cost. The function Q is the *expected* minimum *recourse cost* function or *expected value function*; for any $x \in \mathbb{R}^n$ it describes the expected value of the cost of the recourse. So SLPwR is equivalent to

$$\min_{x \in X}\{cx + Q(x) : Ax = b\}. \tag{3.10}$$

From this representation it is clear that the recourse problem has linear equality constraints, whereas the objective function is nonlinear. Later on we will show that its objective function is convex, actually.

Remark 3.2.2 Sometimes the following representation of SLPwR is appropriate:

$$\min_{x \in X_0} \mathbb{E}_\omega \left[g(\omega, x)\right], \tag{3.11}$$

where $X_0 := \{x \in X : Ax = b\}$ and

$$g(\omega, x) := cx + v(h(\omega) - T(\omega)x), \quad \omega \in \Omega, \ x \in X.$$

This representation focuses on the fact that the model aims at minimizing a so-called expected function (see e.g. Sect. A.1).

Large-Scale Deterministic Equivalent Representation of Recourse Models

From a modeling point of view the following representation of SLPwR is appropriate:

$$
\begin{array}{llll}
(\mathrm{LP}_1(\omega)) \min\limits_{x, y(\omega)} & cx + \mathbb{E}_\omega\,[qy(\omega)] & & \\
\text{s.t.} \quad Ax & = b & & \text{first-stage constraints} \\
T(\omega)x \;+\; Wy(\omega) & \sim h(\omega) & \forall \omega \in \Omega & \text{second-stage constraints} \\
x \in X \qquad y(\omega) \in Y & & & \\
\uparrow \qquad\qquad \uparrow & & & \\
\text{first-stage decisions} \quad \text{second-stage decisions} & & &
\end{array}
$$

It shows, how introduction of the recourse structure is based on a relaxation of the constraints $T(\omega)x \sim h(\omega)$. Notice that there are second-stage constraints for each $\omega \in \Omega$ separately, and that y will depend on ω.

Consider the special case of SLPwR, when Ω is finite. That is, $\Omega = \{\omega^1, \ldots, \omega^S\} \subset \mathbb{R}^r$, and $\Pr\{\omega = \omega^s\} = p_s$, $s = 1, \ldots, S$, describe its known distribution. Then, using the notation $T^s := T(\omega^s)$, $h^s := h(\omega^s)$, it is easy to see that SLPwR is equivalent to the following *large scale linear* programming problem:

$$
\begin{array}{lllllll}
\min\limits_{\substack{x \in \mathbb{R}^n \\ y^s \in \mathbb{R}^p \\ s=1,\ldots,S}} & cx + & p_1 \cdot qy^1 + & p_2 \cdot qy^2 + & \ldots + & p_S \cdot qy^S & \\
& Ax & & & & & = b \\
& T^1 x + & Wy^1 & & & & \sim h^1 \\
& T^2 x & & + \; Wy^2 & & & \sim h^2 \\
& \vdots & & & \ddots & & \\
& T^S x & & & + & Wy^S & \sim h^S \\
& x \in X & y^1 \in Y & y^2 \in Y & & y^S \in Y &
\end{array}
$$

Here, $y^s = y(\omega^s)$ indicates the recourse actions to be taken if the realization ω^s of ω occurs. The advantage of this model is obvious: it is a linear program rather than nonlinear. The only disadvantage is its size: it has $n + pS$ variables and $m_1 + mS$ (explicit) constraints. Usually, it is very large scale, since even with a reasonable number of random variables r, each with a reasonable number of possible realizations, the number S may be astronomical! For instance, 20 random variables with each 10 possible values give rise to $S = 10^{20}$. This is the reason that in practical applications one tries to approximate the given distribution of ω by a *small size* discrete distribution, preferably in such a way that the *optimal value* is approximated as close as possible, rather than the distribution itself (see Sect. 3.3.5).

Canonical Form of Recourse Models

We conclude the list of representations of the model SLPwR with the *canonical form*. Here all constraints are either equality constraints or nonnegativity constraints:

$$\min_{\tilde{x}\geq 0}\left\{\tilde{c}\tilde{x}+\mathbb{E}_{\omega}\left[\min_{\tilde{y}\geq 0}\left\{\tilde{q}\tilde{y}:\tilde{W}\tilde{y}=\tilde{h}(\omega)-\tilde{T}(\omega)\tilde{x}\right\}\right]:\tilde{A}\tilde{x}=\tilde{b}\right\} \tag{3.12}$$

Obviously, such a model is a special case of SLPwR. But any SLPwR can be represented in such a canonical form by including slack variables in the vectors of first and second-stage decisions and adjusting the corresponding matrices and cost vectors. (Constraints like $x^l \leq x \leq x^u$ can be transformed into $x^1 \geq 0, x^2 \geq 0$ and $x^1 + x^2 = x^u - x^l$ and eliminating x via $x = x^l + x^1$.) As in deterministic LP, theory and algorithms are most easily developed for models in canonical form, but users will not have to specify the canonical form themselves; the adjustments are taken care of by the software.

3.2.2 *Modeling Aspects*

Given the 'underlying' LP model, with random constraints, how to transform it into a well-specified SLPwR model? That is, *where does the recourse structure* (Y, q, W) *come from*? There are three possibilities (and convex combinations of them):

(i) *No recourse* actions *are modeled*, but deviations from goals are penalized. Then the recourse variables are more or less artificial: they represent just 'surpluses' or 'shortages' with respect to goals. Examples include simple recourse or multiple simple recourse, where each deviation is penalized separately, but also penalization of the maximum deviation from the goal may be modeled using artificial recourse variables (see the almost simple recourse structure in Exercise 3.1.8).

(ii) Recourse variables are introduced that represent *corrective actions*, to be taken after realization of ω if the goals are not reached. In this case the question comes up, why these additional possible actions were not included in the original model. If there are good reasons for this, one should specify the recourse cost sufficiently high; that means that recourse actions would not be used in the original model if there were no random parameters involved.
More often than not, the underlying model $\text{LP}_0(\omega)$ contains already the variables that can be assigned as recourse variables, see (iii).

(iii) *Split the vector of decision variables* in $\text{LP}_0(\omega)$ in two parts: one set has to be determined before obtaining ω, and the remaining set of variables may depend on the value of ω. The constraints get a corresponding split: constraints not including ω belong to the first stage, the remaining ones to the second stage.

In some cases, this splitting up is rather obvious. For instance, in hierarchical planning models more often than not it is suitable to assume that investment decisions (see the case study "Investment Planning for Electricity Generation" in Sect. 8.6) have to be made 'here-and-now' whereas the actual usage of the investment capacities will be of the wait-and-see type, i.e. depending on the random demand for it. In such models the long-term decisions are first-stage, and the short-term decisions are modeled as second-stage variables.

Similarly, in production planning models (see the case study "Multi-Period Production Planning" in Sect. 8.1) it is not unreasonable to specify production levels as first-stage decisions, whereas inventories and backlogs are typically second-stage variables, with values depending on the actual values of the random demands. Many models are already dynamic, and the time consists of τ stages, say. Then a τ-stage stochastic programming model could be appropriate. Still, also in these circumstances, a two-stage SLP model might be much better than a deterministic model. Usually, the first stage deals with the decisions that have to be taken here-and-now, and (as an approximation) it is assumed that all other decisions can be taken after observation of all uncertain parameters in the τ stages. Together with a 'rolling horizon' procedure, such an approach usually is quite acceptable.

The case studies in Chap. 8 contain several recourse models with different recourse structures. All three possibilities to derive a recourse structure, as discussed above, are present among the case studies.

3.2.3 Special Recourse Structures

Consider the recourse structure (Y, q, W) in relation to the underlying linear programming model $\mathrm{LP}_0(\omega)$. This recourse structure is called *fixed* or deterministic, since its numerical specification is assumed to be known at the first stage. This assumption is not really necessary. It is possible, that the actual recourse possibilities are *random*. Then the recourse model SLPwR remains appropriate, provided the recourse structure is known in the second stage. In that case, just replace (Y, q, W) by $(Y(\omega), q(\omega), W(\omega))$, possibly after extending the random vector ω. For convenience, however, we will restrict the attention to fixed recourse in the sequel.

The question comes up: given $\mathrm{LP}_0(\omega)$, how to select a suitable (fixed) recourse structure? As has been indicated in Sect. 3.2.2, in each special case the model builder will make this selection in accordance to her interpretation of $\mathrm{LP}_0(\omega)$. Here we focus on general theoretical observations. Consider the recourse penalty function v, given in (3.8). It represents the optimal value of the LP problem determined by the recourse structure, as a function of the right-hand side vector z. In principle it is possible that for some values of z, the value $v(z) = +\infty$: then the LP problem is infeasible. Or it is possible that for some z, the value $v(z) = -\infty$: then the LP problem is unbounded. In the first case there may exist deviations $h(\omega) - T(\omega)x$ that cannot be corrected in the second stage of SLPwR. In the second case there may exist deviations $h(\omega) - T(\omega)x$ that are extremely advantageous. Both situations should be avoided. In the first case, the recourse structure is not rich enough to cover all possible deviations, whereas in the second case the penalization of deviations has failed completely.

Definition 3.2.3 The fixed recourse structure (Y, q, W) is called *relatively complete* (i.e. relative with respect to $\mathrm{LP}_0(\omega)$) if for all $\omega \in \Omega$ and all $x \in X$ with $Ax = b$ there exists a $y \in Y$, possibly depending on ω and x, such that $Wy \sim h(\omega) - T(\omega)x$. Equivalently, if $v(h(\omega) - T(\omega)x) < +\infty$ for all $\omega \in \Omega$ and $x \in \{x \in X : Ax = b\}$.

Definition 3.2.4 The fixed recourse structure (Y, q, W) is called *complete* if for all $z \in \mathbb{R}^m$ there exists a $y \in Y$, possibly depending on z, such that $Wy \sim z$. Equivalently, if $v(z) < +\infty$ for all $z \in \mathbb{R}^m$.

Notice that completeness implies relative completeness, but in general the reverse implication is not true. Both concepts do not depend on q, actually. In the canonical form (see (3.12)), completeness is determined by the recourse matrix $\tilde{W}$ *only*. Such a matrix is called a *complete recourse matrix* (crm). In applications one chooses a $\tilde{W}$ that is crm. If not, there is no problem if the recourse structure is relatively complete (but in general this condition is not easy to verify). If the recourse is not relatively complete, the restricted recourse possibilities give rise to additional constraints on the first-stage decision variables x

$$\Pr\{\omega \in \Omega : v(h(\omega) - T(\omega)x) = +\infty\} = 0.$$

Such so-called *induced constraints* are not easy to detect, although some solution procedures are able to cope with them.

Exercise 3.2.5 Give a dual characterization of crm $\tilde{W}$; see Assignment R2 (a). ◁

Definition 3.2.6 The fixed recourse structure (Y, q, W) is called *extremely inexpensive* if there exists a $z \in \mathbb{R}^m$ such that $v(z) = -\infty$.

Obviously, extremely inexpensive recourse structures are not suitable. Usually, they are avoided by assuming $q \geq 0$.

Remark 3.2.7 Sometimes negative penalties are appropriate. For instance, if the random constraints are $T(\omega)x \geq h(\omega)$, it might be suitable to reward surpluses $\big(T(\omega)x - h(\omega)\big)^+$ (while penalizing shortages $\big(T(\omega)x - h(\omega)\big)^-$).

Exercise 3.2.8 Give a dual characterization of not extremely inexpensive canonical recourse structures; see Assignment R2 (b). ◁

Definition 3.2.9 A fixed recourse structure is called *simple recourse* if $\tilde{y} = (y^+, y^-) \in Y = \mathbb{R}_+^{2m}$, and up to column and/or row permutations of $\tilde{W}$, its canonical form can be written as

$$\begin{pmatrix} \tilde{q} \\ \tilde{W} \end{pmatrix} = \begin{pmatrix} q^+ & q^- \\ I & -I \end{pmatrix}, \tag{3.13}$$

where q^+, q^- are $1 \times m$ unit cost vectors and I denotes the $m \times m$ identity matrix.

Exercise 3.2.10 Show that $\tilde{W}$ is a crm, and that the simple recourse structure is not extremely inexpensive if and only if $q^+ + q^- \geq 0$. ◁

The use of simple recourse means, that the deviations in each goal constraint $t_i(\omega)x \sim h_i$ are penalized separately and linearly, where 'surpluses and 'shortages' have different unit costs. Similar remarks apply to the multiple simple recourse structure in Exercise 3.1.7.

The main reason to call the structure (3.13) 'simple' is the fact that it is possible to solve all second-stage LP's in closed form. Indeed, by denoting the recourse variables corresponding to the first part of $\tilde{W}$ by y^+, and those of the second part by y^- (with unit penalty cost vectors q^+ and q^-, respectively, with $q^+ + q^- \geq 0$) one finds for the minimal *simple* recourse penalty cost function $v(z)$, $z \in \mathbb{R}^m$,

$$v(z) = \min_{y^+ \geq 0, y^- \geq 0} \left\{ q^+ y^+ + q^- y^- : y^+ - y^- = z \right\} \tag{3.14}$$

$$= \sum_{i=1}^{m} \min_{y_i^+ \geq 0, y_i^- \geq 0} \left\{ q_i^+ y_i^+ + q_i^- y_i^- : y_i^+ - y_i^- = z_i \right\}$$

$$= \sum_{i=1}^{m} v_i(z_i) \tag{3.15}$$

with $v_i(z_i) = q_i^+ (z_i)^+ + q_i^- (z_i)^-$, where $(z_i)^+$ (c.q. $(z_i)^-$) denotes the positive (negative) part of the ith component of z.

Exercise 3.2.11 Suppose that for some i, $q_i^+ + q_i^- = 0$. Then, $v_i(z_i) = q_i^+ (z_i)^+ + (-q_i^+)(z_i)^- = q_i^+ z_i$ is linear on $\mathbb{R}$. Is this a suitable penalty function? ◁

Simple recourse models are popular for two reasons. First of all, they can be solved much easier than general recourse models, as will appear later. Secondly, it describes a simple but still rather flexible way to deal with random constraints in decision problems: just penalize each individual surplus or shortage in $t_i(\omega)x \sim h_i(\omega)$ separately and linearly. On the other hand, the name is somewhat confusing. More often than not in fact one does not aim at recourse actions when applying simple recourse. Moreover, there are other recourse structures that are also easy to deal with (e.g. multiple simple recourse, see Exercise 3.1.7, and almost simple recourse, see Exercise 3.1.8).

3.3 Properties of Recourse Models

Properties of recourse models obviously depend on properties of the expected minimum recourse cost function Q, defined in (3.9). To study properties of this function, we first need to define what we mean by the expectation of a function defined on $\mathbb{R}^r \times \mathbb{R}^n$, where $\mathbb{R}^r$ contains the support Ω of a random vector ω.

Let ω be a random vector with support $\Omega \subset \mathbb{R}^r$ and cumulative distribution function (cdf) F. The expectation with respect to ω of a function $g(\omega, x)$ from $\mathbb{R}^r \times \mathbb{R}^n$ to $[-\infty, \infty]$ is given by

$$\mathbb{E}_\omega [g(\omega, x)] = \int_\Omega g(t, x)\, dF(t), \quad x \in \mathbb{R}^n, \tag{3.16}$$

where the right-hand side is a so-called Riemann–Stieltjes integral. Appendix A contains a discussion of the meaning of (3.16) and several useful theorems for expected functions.

3.3.1 Fixed Recourse

Consider the recourse model SLPwR in canonical form:

$$\text{(SLPwR)} \quad \min_{x \geq 0} \left\{ cx + \mathbb{E}_\omega \left[\min_{y \geq 0} \{qy : Wy = h(\omega) - T(\omega)x\} \right] : Ax = b \right\},$$

where the $m \times n$ matrix $T(\omega)$ and $m \times 1$ vector $h(\omega)$ both depend on a random vector $\omega \in \mathbb{R}^r$. In principle it is possible that the whole of T and h is random. Then $r := m(n+1)$ is appropriate. But more often than not only a restricted number of matrix and vector elements are random, so that $r < m(n+1)$. Moreover, without loss of generality we may (and will) assume, that $T(\omega)$ and $h(\omega)$ are *linear* functions of $\omega = (\omega_1, \ldots, \omega_r)$. Thus,

$$T(\omega) \equiv T_0 + \sum_{i=1}^{r} \omega_i T_i,$$

$$h(\omega) \equiv h_0 + \sum_{i=1}^{r} \omega_i h_i,$$

where for each $i = 0, 1, \ldots, r$, T_i is a known $m \times n$ matrix and h_i is a known $m \times 1$ vector.

Theorem 3.3.1 illustrates the most important properties of the expected value function

$$Q(x) = \mathbb{E}_\omega \left[\min_{y \geq 0} \{qy : Wy = h(\omega) - T(\omega)x\} \right], \quad x \in \mathbb{R}^n.$$

Theorem 3.3.1 *Consider a SLPwR model with fixed recourse structure* (Y, q, W). *If*

(a) the recourse structure is complete,
(b) the recourse structure is not extremely inexpensive,
(c) $\mathbb{E}_\omega [|\omega_i|] < \infty$, $i = 1, \ldots, r$,

then $Q(x)$, $x \in \mathbb{R}^n$ *is a finite convex function. Moreover, it is polyhedral (piecewise linear) if the distribution of* ω *is discrete, and it is everywhere differentiable if the distribution of* ω *is a continuous one.*

Proof The proof of this theorem relies on properties of the recourse penalty function v, defined as

$$v(z) = \inf_{y \in \mathbb{R}^p} \{qy : Wy = z, \ y \geq 0\}, \quad z \in \mathbb{R}^m.$$

From the point of view of deterministic LP, the function v is just the optimal value function of the second-stage LP problem, as a function of the right-hand side vector $z \in \mathbb{R}^m$. Properties of this function are presented in Appendix C. Under conditions (a) and (b) above, it holds that v is a finite convex polyhedral function, given by

$$v(z) = \max_{k=1,\ldots,K} \lambda^k z, \quad z \in \mathbb{R}^m, \tag{3.17}$$

where λ^k, $k = 1, \ldots, K$, are the extreme points of the nonempty bounded set $\Lambda = \{\lambda \in \mathbb{R}^m : \lambda W \leq q\}$.

Now we will indicate why condition (c) implies that $Q(x)$ *is finite for all* $x \in \mathbb{R}^n$, given that v is finite for all $z \in \mathbb{R}^m$. Fix $x \in \mathbb{R}^n$. Since we assumed that $h(\omega)$ and $T(\omega)$ depend *linearly* on $\omega \in \mathbb{R}^r$, there exists a fixed vector $a \in \mathbb{R}^m$ and a fixed $m \times r$ matrix B (both depending on x of course) such that

$$h(\omega) - T(\omega)x = a - B\omega, \quad \omega \in \Omega \subset \mathbb{R}^r.$$

Exercise 3.3.2 Verify this. ◁

Defining the scalars

$$\alpha_k := \lambda^k a, \quad k = 1, \ldots, K,$$

and the row vectors in $\mathbb{R}^r$

$$\beta_k := \lambda^k B, \quad k = 1, \dots, K,$$

it follows from (3.9) and (3.17) that

$$Q(x) = \mathbb{E}_\omega \left[\max_{k=1,\dots,K} (\alpha_k - \beta_k \omega) \right].$$

From this representation it is not difficult to show that $Q(x)$ is finite, if $E|\omega_i| < \infty$ for all $i = 1, \dots, r$.

Exercise 3.3.3 Show this. ◁

Next we indicate why Q is a *convex* function, actually. This follows directly from its definition (3.9). Since v is a convex function, and $h(\omega) - T(\omega)x$ is linear in x for any $\omega \in \Omega$, we have that for any fixed $\omega \in \Omega$, the function

$$g(\omega, x) := v(h(\omega) - T(\omega)x), \quad x \in \mathbb{R}^n,$$

is convex in x too (see Appendix B). Since the mean value of a collection of convex functions is a convex function again (see Appendix A), it follows that

$$Q(x) = \mathbb{E}_\omega \left[g(\omega, x) \right], \quad x \in \mathbb{R}^n,$$

is convex, indeed. Moreover, if ω has a discrete distribution, say $\Pr\{\omega = \omega^s\} = p_s$, $s = 1, \dots, S$, then the convex function

$$Q(x) = \sum_{s=1}^{S} p_s \cdot v(h(\omega^s) - T(\omega^s)x), \quad x \in \mathbb{R}^n,$$

is even piecewise linear in x. The number of linear pieces may be very large, though.

Exercise 3.3.4 Show that for x in such a piece, one has

$$Q(x) = \sum_{s=1}^{S} p_s \cdot \lambda^{k(x,s)} \left(h(\omega^s) - T(\omega^s)x \right).$$

How to characterize $\lambda^{k(x,s)}$? ◁

It follows that

$$\nabla Q(x^0) = \sum_{s=1}^{S} p_s \cdot \lambda^{k(x^0,s)} \left(-T(\omega^s) \right),$$

at least if $\lambda^{k(x,s)}$ is constant for x in the neighborhood of x^0.

Finally, we state that using convex analysis one may derive a formula for the subdifferential of Q at x, by integrating the subdifferential of v and using the chain rule:

$$\partial Q(x) = \mathbb{E}_\omega \left[\partial_x v(h(\omega) - T(\omega)x)\right] = \left\{\mathbb{E}_\omega \left[-\lambda^{(x,\omega)} T(\omega)\right]\right\},$$

where $\lambda^{(x,\omega)}$ denotes any optimal solution of the dual problem $\sup_{\lambda \in \mathbb{R}^m} \{\lambda(h(\omega) - T(\omega)x) : \lambda W \leq q\}$ (see Appendix C for a characterization of the subdifferential $\partial v(z)$).

□

3.3.2 *Simple Recourse*

Now consider the fixed recourse model in canonical form under the assumption that the recourse is simple (see Sect. 3.1):

$$\begin{pmatrix} q \\ W \end{pmatrix} := \begin{pmatrix} q^+ & q^- \\ I & -I \end{pmatrix}, \qquad Y = \left\{y = (y^+, y^-) \in \mathbb{R}^{2m} : y \geq 0\right\},$$

where q^+, q^- are $1 \times m$ vectors and I is the $m \times m$ identity matrix. Obviously, this recourse structure is complete, and it is not extremely inexpensive if $q^+ + q^- \geq 0$ (Exercise 3.2.10). Therefore, we assume $q^+ + q^- \geq 0$, and also $E|\omega_i| < \infty$, $i = 1, \ldots, r$. Then all conditions of Theorem 3.3.1 are satisfied. In this case, however, we have a simple representation of v, and therefore of Q, that we will derive now. From (3.15) at the end of Sect. 3.2.3 we see that the simple recourse model can be written as

$$\min_{x \in \mathbb{R}^n} \left\{ cx + \mathbb{E}_\omega \left[\sum_{i=1}^m v_i(h_i(\omega) - T_i(\omega)x) \right] : Ax = b,\ x \geq 0 \right\},$$

where $v_i(z_i) = q_i^+(z_i)^+ + q_i^-(z_i)^-$, $z_i \in \mathbb{R}$, and where $h_i(\omega)$ and $T_i(\omega)$ denote the ith rows of $h(\omega)$ and $T(\omega)$, respectively. Hence an equivalent formulation is

$$\min_{x \in \mathbb{R}^n} \left\{ cx + \sum_{i=1}^m \left(q_i^+ \mathbb{E}_\omega \left[\left(h_i(\omega) - T_i(\omega)x\right)^+ \right] + q_i^- \mathbb{E}_\omega \left[\left(h_i(\omega) - T_i(\omega)x\right)^- \right] \right) : \right.$$
$$\left. Ax = b,\ x \geq 0 \right\},$$

where for any $z \in \mathbb{R}^m$ we denote the vector of positive (negative) parts of its components by $(z)^+$ ($(z)^-$, respectively). By 'differentiating under the integral' (see Appendix A) one may get explicit formulas for $Q(x)$ and its subdifferential $\partial Q(x)$ (or gradient $\nabla Q(x)$, if existing).

We will work out some details for an important special case: *only the right-hand side is random*. That is

$$h(\omega) := \omega \in \mathbb{R}^m$$
$$T(\omega) := T \quad \text{(deterministic)}$$

Then the simple recourse model in canonical form reduces to

$$\min_{x \in \mathbb{R}^n} \left\{ cx + \sum_{i=1}^{m} \mathbb{E}_{\omega_i}\left[v_i(\omega_i - T_i x) \right] : Ax = b,\ x \geq 0 \right\}.$$

Notice that only the marginal distributions of each ω_i are relevant here (Interpret this phenomenon). By introducing additional first-stage decision variables $s \in \mathbb{R}^m$, with components s_i, $i = 1, \ldots, m$, (called *tender* variables) we get

$$\min_{\substack{x \in \mathbb{R}^n \\ s \in \mathbb{R}^m}} \left\{ cx + \sum_{i=1}^{m} Q_i(s_i) : \begin{array}{rl} Ax & = b \\ Tx - s & = 0 \\ x \geq 0 & \end{array} \right\}$$

where

$$\begin{aligned} Q_i(s_i) &:= \mathbb{E}_{\omega_i}\left[v_i(\omega_i - s_i) \right] \\ &= q_i^+ \mathbb{E}_{\omega_i}\left[(\omega_i - s_i)^+ \right] + q_i^- \mathbb{E}_{\omega_i}\left[(\omega_i - s_i)^- \right], \quad s_i \in \mathbb{R}. \end{aligned} \tag{3.18}$$

The introduction of the tender variables has the advantage, that the nonlinear (convex) part in the objective function is a *separable* function: a sum of functions of one variable each. Therefore *separable convex programming algorithms can be used to solve simple recourse problems with only the right-hand side random*, at least if we are able to provide formulas for $Q_i(s_i)$ and $Q_i'(s_i)$ (c.q. $\partial Q_i(s_i)$). This is the case, indeed, as we will indicate now. The properties of the so-called *one-dimensional simple recourse function* Q_i follow directly from those of the functions

$$G_i(s_i) := \mathbb{E}_{\omega_i}\left[(\omega_i - s_i)^+ \right], s_i \in \mathbb{R} : \text{ the } \textit{expected surplus function}, \tag{3.19}$$

$$H_i(s_i) := \mathbb{E}_{\omega_i}\left[(\omega_i - s_i)^- \right], s_i \in \mathbb{R} : \text{ the } \textit{expected shortage function}, \tag{3.20}$$

see (3.18). The properties of the functions G_i, H_i, and Q_i, which depend on the distribution of ω_i, have been worked out below.

We conclude this exposition by considering the special case where, in addition to the assumptions made above, it also is assumed that each ω_i has a finite discrete distribution. Then the simple recourse model with only right-hand side random can be solved as an LP problem, whose size is not too large in relation to the size of the underlying LP model.

Exercise 3.3.5 Verify this; see Assignment R3. Show that its size is much smaller than that of the 'brute force' model of Sect. 3.2.1. Explain this. ◁

3.3.3 Expected Shortage and Surplus Functions

In this section we deal with two one-dimensional functions that play an important role in Stochastic Programming. To derive their properties we will apply several results and concepts introduced above.

Definition 3.3.6 Let ω be a one-dimensional random variable. Then the *expected shortage function* is

$$H(x) = \mathbb{E}_\omega\left[(\omega - x)^-\right], \qquad x \in \mathbb{R}.$$

Definition 3.3.7 Let ω be a one-dimensional random variable. Then the *expected surplus function* is

$$G(x) = \mathbb{E}_\omega\left[(\omega - x)^+\right], \qquad x \in \mathbb{R}.$$

The name of the function H stems from the following. Let the random variable ω represent the uncertain demand for some product. If we interpret x as the supply of this product, then the amount $H(x)$ is the expected shortage of demand relative to supply. Indeed, if we choose to supply x then for each realization $\bar{\omega}$ the resulting shortage (if any) is the maximum of $(x - \bar{\omega})$ and 0. As before, we use the notation $(\bar{\omega} - x)^-$ to denote this maximum. Taking the expectation over all possible realizations of the random variable ω gives $H(x)$. Similarly, the quantity $G(x)$ can be interpreted as the expected surplus of the uncertain demand ω over supply x.

Remark 3.3.8 The terminology may be a little bit confusing. Since $(\omega - x)^- = (x - \omega)^+$, the shortage of demand relative to supply can also be interpreted as the surplus relative to the demand. Therefore, $H(x)$ is not only the expected demand shortage but also the expected supply surplus. Similarly, $G(x)$ denotes not only the expected demand surplus but also the expected supply shortage.

The functions H and G determine each other completely, as the next exercise shows.

Exercise 3.3.9 Show that $H(x) - G(x) = x - \mathbb{E}_\omega[\omega]$ if $\mathbb{E}_\omega[\omega]$ is finite. ◁

We give explicit formulas for the functions H and G below. These formulas can be interpreted as integrals over the tails of the distribution function F and the complementary distribution function $1-F$. As we will see, many properties of these functions can be derived from these expressions.

$$H(x) = \int_{-\infty}^{x} F(t)\,dt \tag{3.21}$$

$$G(x) = \int_x^{\infty} \big(1 - F(t)\big)\, dt \tag{3.22}$$

Here F denotes the cdf of ω. Equivalent representations are

$$H(x) = \int_{-\infty}^{x} \Pr\{\omega \leq t\}\, dt = \int_{-\infty}^{x} \Pr\{\omega < t\}\, dt$$

and

$$G(x) = \int_x^{\infty} \Pr\{\omega \geq t\}\, dt = \int_x^{\infty} \Pr\{\omega > t\}\, dt.$$

We will only prove equality (3.21), the proof of (3.22) is similar and left as an exercise.

$$\begin{aligned} H(x) &= \mathbb{E}_{\omega}\left[(\omega - x)^{-}\right] \\ &= \int_{-\infty}^{\infty} (s - x)^{-}\, dF(s) \\ &= \int_{-\infty}^{x} (x - s)\, dF(s). \end{aligned} \tag{3.23}$$

Using $x - s = \int_s^x dt$ and then changing the order of integration we obtain

$$\begin{aligned} H(x) &= \int_{-\infty}^{x} \int_s^x dt\, dF(s) \\ &= \int_{-\infty}^{x} \int_{-\infty}^{t} dF(s)\, dt \\ &= \int_{-\infty}^{x} F(t)\, dt. \end{aligned}$$

Expected Shortage Function H

Next we present a list of properties of the function H.

Theorem 3.3.10 *Let ω be a random variable with cdf F. Define the expected shortage function H as*

$$H(x) = \mathbb{E}_{\omega}\left[(\omega - x)^{-}\right], \qquad x \in \mathbb{R}.$$

Then

(a) *H is non-negative.*
(b) *H is non-decreasing.*
(c) *H is convex.*
(d) *For all $x \in \mathbb{R}$, $H(x) < \infty$ if and only if $\mu^- = \mathbb{E}_\omega\left[(\omega)^-\right] < \infty$.*

Under the assumption that $\mu^- < \infty$:

(e) *H is continuous.*
(f) *H is Lipschitz continuous with constant 1.*
(g) *The left and right derivative of H exist everywhere and are given by $H'_-(x) = \Pr\{\omega < x\}$ and $H'_+(x) = \Pr\{\omega \leq x\}$, respectively.*
(h) *H is subdifferentiable with subdifferential*

$$\partial H(x) = [\Pr\{\omega < x\}, \Pr\{\omega \leq x\}].$$

(i) *H is differentiable in any continuity point x_0 of F with derivative $H'(x_0) = F(x_0)$.*
(j) *The curve $y = H(x)$ has a horizontal asymptote 0 at $-\infty$.*

If also $\mu^+ = E[(\omega)^+] < \infty$, then

(k) *The curve $y = H(x)$ has $x - \mu$ as asymptote at $+\infty$, where $\mu = \mu^+ - \mu^- = \mathbb{E}_\omega[\omega]$.*
(l) *Outside the convex hull of the support of ω the curve $y = H(x)$ coincides with its asymptotes.*

Proof

(a) and (b) are trivial.
(c) By Theorem A.1.1.
(d) Using (D.1) in Appendix D, it follows that

$$\mu^- - |x| \leq H(x) \leq \mu^- + |x| \qquad \forall x \in \mathbb{R}.$$

(e) Follows from (f).
(f) For all x and y in $\mathbb{R}$,

$$\left|(t-x)^- - (t-y)^-\right| \leq |x-y| \qquad \forall t \in \mathbb{R},$$

so that

$$\begin{aligned}\left|\mathbb{E}_\omega\left[(\omega-x)^-\right] - \mathbb{E}_\omega\left[(\omega-y)^-\right]\right| &\leq \mathbb{E}_\omega\left[\left|(\omega-x)^- - (\omega-y)^-\right|\right] \\ &\leq |x-y|.\end{aligned}$$

(g) Applying Theorem A.1.6 (see Example A.1.7) or, directly from the definition of the right derivative and (3.21):

$$H'_+(x) = \lim_{y \downarrow x} \frac{H(y) - H(x)}{y - x}$$
$$= \lim_{y \downarrow x} \frac{\int_x^y F(t)\,dt}{y - x}$$
$$= \lim_{\varepsilon \downarrow 0} \frac{\int_0^\varepsilon F(x+t)\,dt}{\varepsilon}.$$

Applying L'Hôpital's rule we obtain

$$H'_+(x) = \lim_{\varepsilon \downarrow 0} F(x+\varepsilon) = \Pr\{\omega \le x\}.$$

Similarly, $H'_-(x) = \Pr\{\omega < x\}$.

(h) By Theorem B.30 (iv) we have

$$\partial H(x) = [H'_-(x), H'_+(x)] \qquad \forall x \in \mathbb{R}.$$

Note that if the cdf F has a discontinuity at x (i.e., $\Pr\{\omega = x\} > 0$) then H has a 'knot' at x.

(i) Trivial, since in that case $H'_-(x) = H'_+(x)$.

(j) Since $H(x)$ is finite for all $x \in \mathbb{R}$, it holds that

$$\lim_{x \to -\infty} H(x) = \lim_{x \to -\infty} \int_{-\infty}^x F(t)dt = 0.$$

(k) Since $H(x) - (x - \mu) = G(x)$, see Exercise 3.3.9, we have to show that $\lim_{x\to\infty} G(x) = 0$. The proof of this is similar to that of (j). First, it is shown that (3.22) is true for all x; then it is shown that $G(x)$ is finite for all x if $\mu^+ < \infty$; and finally

$$\lim_{x \to \infty} G(x) = \lim_{x \to \infty} \int_x^\infty (1 - F(t))dt = 0.$$

(l) Suppose $\Pr\{\omega \ge a\} = 1$ for some $a > -\infty$. Then we have $\Pr\left\{(\omega - x)^- = 0\right\} = 1$ for all $x \le a$, so that $H(x) = 0$ for $x \le a$. If $\Pr\{\omega \le b\} = 1$ for some $b < \infty$, then $\Pr\left\{(\omega - x)^- = x - \omega\right\} = 1$ for all $x \ge b$, so that $H(x) = \mathbb{E}_\omega[(x - \omega)] = x - \mu$ for all $x \ge b$.

□

See Figs. 3.1 and 3.2 for the graph of the function H based on a continuous and a discrete distribution of ω, respectively.

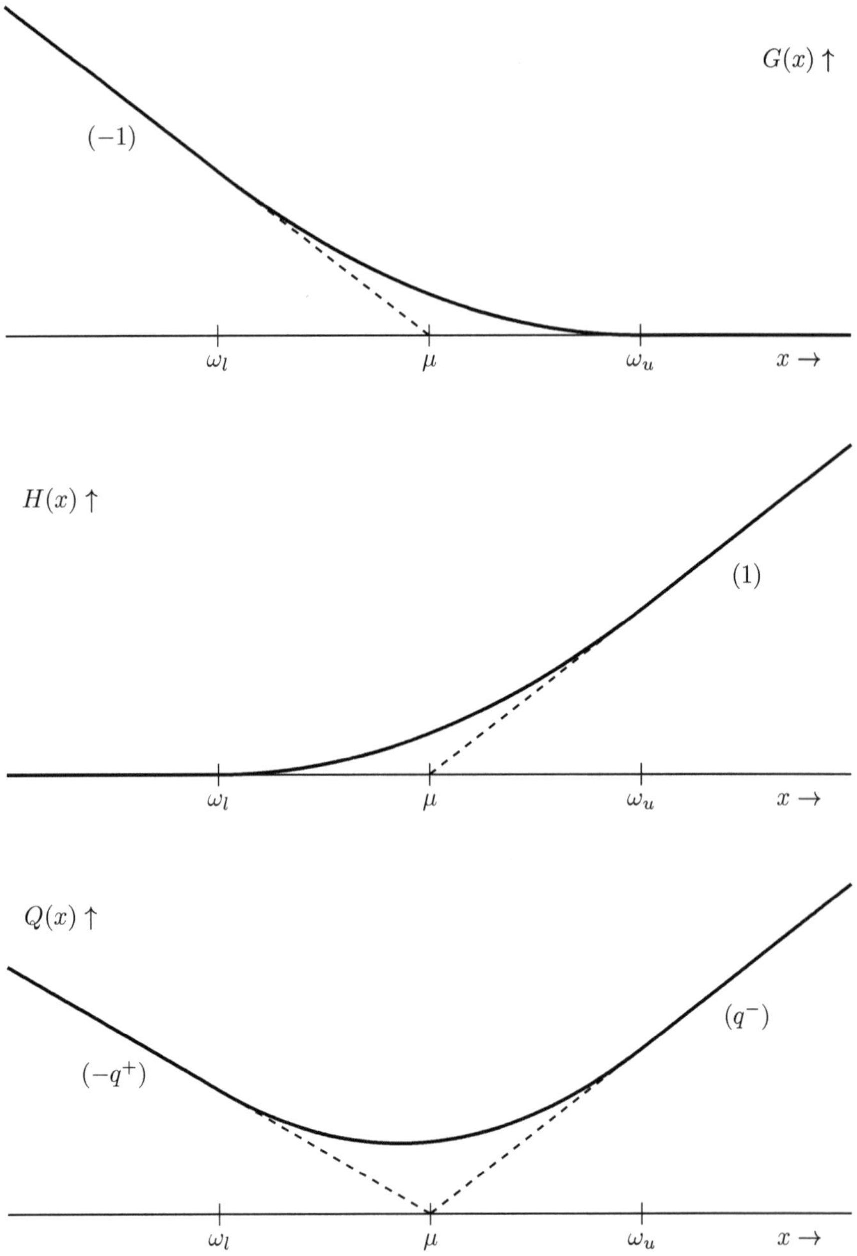

Fig. 3.1 The functions G, H and Q and their asymptotes. Here, $\Omega \subset [\omega_l, \omega_u]$ and $\mathbb{E}_\omega[\omega] = \mu$. In fact, the distribution of ω is uniform on $[\omega_l, \omega_u]$

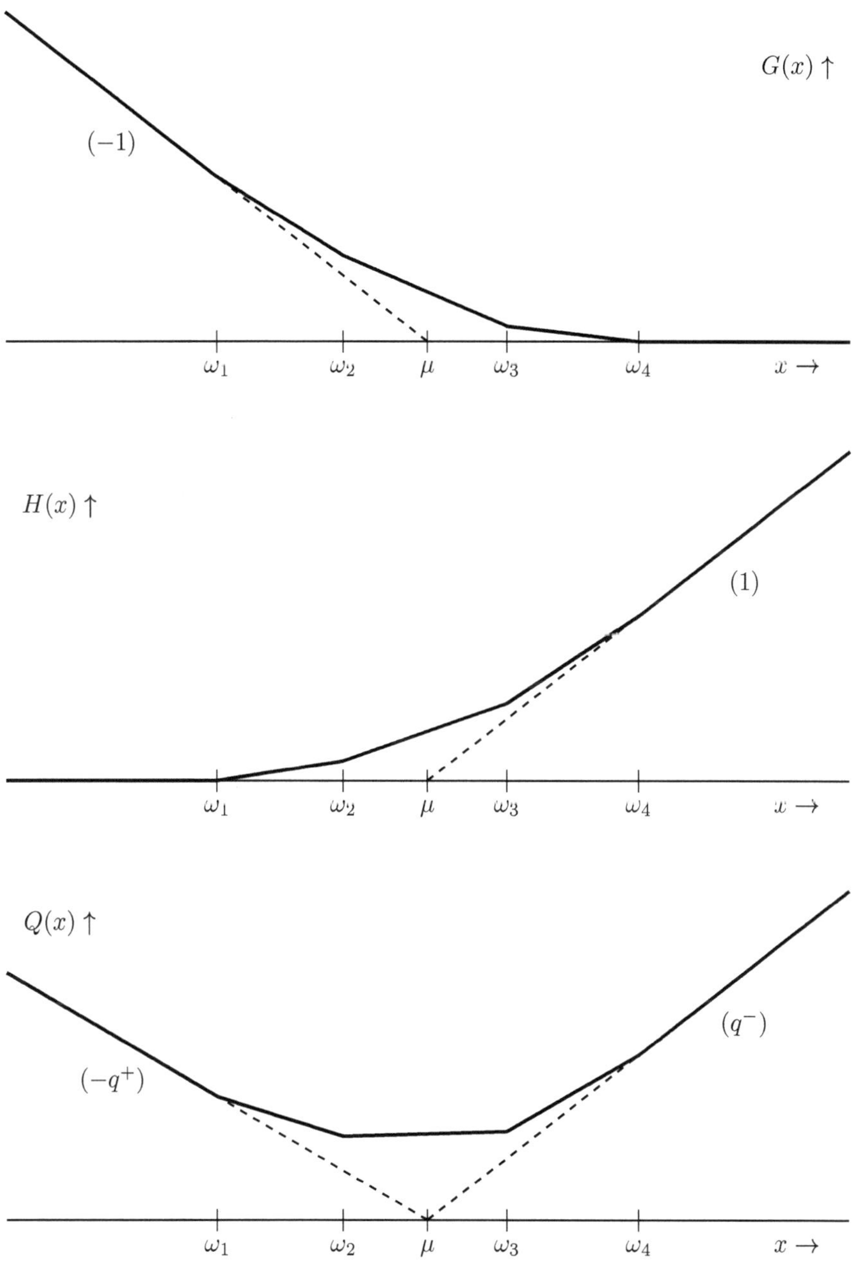

Fig. 3.2 The functions G, H and Q and their asymptotes. Here, $\Pr\{\omega = \omega_i\} = p_i$ with $p = (0.2, 0.25, 0.4, 0.15)$, and $\mathbb{E}_\omega[\omega] = \mu$

Exercise 3.3.11 Show that

$$H(x) = \sigma^2 f(x) - (\mu - x)F(x), \quad x \in \mathbb{R},$$

with f the pdf and F the cdf of ω, if ω follows a normal distribution with mean μ and variance σ^2.
Hint: use that, for all $s \in \mathbb{R}$, $sf(s) = \mu f(s) - \sigma^2 f'(s)$ in this case. ◁

Expected Surplus Function G

Now we present a similar list of properties of the function G. They can be shown in the same way as has been done for H, or, more easily, follow from these using the result of Exercise 3.3.9.

Theorem 3.3.12 *Let ω be a random variable with cdf F. Define the expected surplus function G as*

$$G(x) = \mathbb{E}_\omega\left[(\omega - x)^+\right], \quad x \in \mathbb{R}.$$

Then

(a) G is non-negative.
(b) G is non-increasing.
(c) G is convex.
(d) For all $x \in \mathbb{R}$, $G(x) < \infty$ if and only if $\mu^+ = \mathbb{E}_\omega\left[(\omega)^+\right] < \infty$.

Under the assumption that $\mu^+ < \infty$:

(e) G is continuous.
(f) G is Lipschitz continuous with constant 1.
(g) The left and right derivative of G exist everywhere and are given by $G'_-(x) = -\Pr\{\omega \geq x\}$ and $G'_+(x) = -\Pr\{\omega > x\}$, respectively.
(h) G is subdifferentiable with subdifferential

$$\partial G(x) = [-\Pr\{\omega \geq x\}, \; -\Pr\{\omega > x\}].$$

(i) G is differentiable in any continuity point x_0 of F, with derivative $G'(x_0) = F(x_0) - 1$.
(j) The curve $y = G(x)$ has a horizontal asymptote 0 at $+\infty$.

If also $\mu^- = \mathbb{E}_\omega\left[(\omega)^-\right] < \infty$, then

(k) The curve $y = G(x)$ has $\mu - x$ as asymptote at $-\infty$, where $\mu = \mu^+ - \mu^- = \mathbb{E}_\omega[\omega]$.
(l) Outside the convex hull of the support of ω the curve $y = G(x)$ coincides with its asymptotes.

Exercise 3.3.13 Prove Theorem 3.3.12 directly, i.e., without reference to Theorem 3.3.10 (except in part (k)). ◁

See Figs. 3.1 and 3.2 for the graph of the function G based on a continuous and a discrete distribution of ω, respectively.

Expected Value Function Q

The properties of the functions G and H can be used to characterize the so-called *one-dimensional expected optimal value function* Q, defined by

$$Q(x) = \mathbb{E}_\omega \left[\inf_{y_1, y_2} \{q_1 y_1 + q_2 y_2 : y_1 - y_2 = \omega - x;\ y_1, y_2 \in \mathbb{R}_+\} \right], \quad x \in \mathbb{R},$$

where ω is a random variable with cdf F and finite mean value μ, and q_1 and q_2 are two scalars. That is, Q is the expected function of the optimal value of a very simple linear programming problem. If $q_1 + q_2 < 0$, we have $Q(x) = -\infty$ for all x (take $y_1 = (\omega - x)^+ + u$, $y_2 = (\omega - x)^- + u$ and let $u \to \infty$). Leaving out this uninteresting case, we *assume that* $q_1 + q_2 \geq 0$. If $q_1 + q_2 = 0$, then $Q(x) = q_1(\mu - x)$, so only the case $q_1 + q_2 > 0$ is left to be analyzed. In this case the linear program has a unique optimal solution

$$y_1^\star = (\omega - x)^+, \quad y_2^\star = (\omega - x)^-$$

so that

$$\begin{aligned} Q(x) &= \mathbb{E}_\omega \left[q_1(\omega - x)^+ + q_2(\omega - x)^- \right] \\ &= q_1 \cdot G(x) + q_2 \cdot H(x), \quad x \in \mathbb{R}. \end{aligned}$$

Hence, all properties of Q follow from those of G and H. In fact, since $H(x) - G(x) = x - \mu$, $x \in \mathbb{R}$, they follow from each of them:

$$Q(x) = (-q_1)(x - \mu) + (q_1 + q_2)H(x), \quad x \in \mathbb{R},$$

$$Q(x) = (q_2)(x - \mu) + (q_1 + q_2)G(x), \quad x \in \mathbb{R}.$$

Remark 3.3.14 Since q_1 represents the cost of surplus per unit, it usually is denoted by q^+. Similarly, since q_2 represents the cost of shortage per unit, it is usually denoted by q^-. This custom is a little bit confusing, since there is no $q \in \mathbb{R}$ such that $q^+ = (q)^+ := \max(0, q)$ and $q^- = (q)^- := \max(0, -q)$. Indeed, usually $q^+ > 0$ and $q^- > 0$. Similarly, the variables y_1 and y_2 are denoted as y^+ and y^-, reflecting the fact that their *optimal values* describe surplus and the shortage, respectively. But only at the optimum one has $y^+ \cdot y^- = 0$!

We conclude this section by giving a list of properties of the one-dimensional expected (optimal) value function. They follow directly from Theorems 3.3.10 and 3.3.12.

Theorem 3.3.15 *Let the one-dimensional expected value function Q be defined as*

$$Q(x) = \mathbb{E}_\omega \left[\min_{y^+, y^-} \{q^+ y^+ + q^- y^- : y^+ - y^- = \omega - x;\ y^+, y^- \in \mathbb{R}_+\} \right], \qquad x \in \mathbb{R},$$

where ω is a random variable with cdf F and finite mean value μ, and where q^+ and q^- are scalars satisfying $q^+ + q^- \geq 0$. Then, for all $x \in \mathbb{R}$,

$$\begin{aligned} Q(x) &= \mathbb{E}_\omega \left[q^+(\omega - x)^+ + q^-(\omega - x)^- \right] \\ &= (-q^+)(x - \mu) + (q^+ + q^-) \int_{-\infty}^{x} F(t)\, dt \\ &= (q^-)(x - \mu) + (q^+ + q^-) \int_{x}^{\infty} (1 - F(t))\, dt. \end{aligned}$$

The function Q satisfies the following properties.

(a) Q is finite, convex, continuous, even Lipschitz continuous with constant $\max\{|q^+|, |q^-|\}$.
(b) The left and right derivatives of Q exist everywhere and are given by

$$Q'_-(x) = -q^+ + (q^+ + q^-) \Pr\{\omega < x\}, \qquad x \in \mathbb{R},$$

$$Q'_+(x) = -q^+ + (q^+ + q^-) \Pr\{\omega \leq x\}, \qquad x \in \mathbb{R}.$$

(c) Q is subdifferentiable with subdifferential $\partial Q(x) = [Q'_-(x), Q'_+(x)]$, for any $x \in \mathbb{R}$. If $\Pr\{\omega = x_0\} = 0$ then Q is differentiable at $x = x_0$, with derivative $Q'(x_0) = -q^+ + (q^+ + q^-)F(x_0)$.
(d) The curve $y = Q(x)$ has the asymptote $q^+(\mu - x)$ at $-\infty$ and the asymptote $q^-(x - \mu)$ at $+\infty$. Outside the convex hull of the support of ω the curve $y = Q(x)$ coincides with the asymptotes.

Exercise 3.3.16 Prove the Lipschitz constant in Theorem 3.3.15 (a). ◁

See Figs. 3.1 and 3.2 for the graph of the function Q based on a continuous and a discrete distribution of ω, respectively. Which are the changes when the support of ω becomes unbounded? What is the graph of Q if $q^+ < 0$ (but $q^- + q^+ \geq 0$)?

3.3.4 Bounds on the Optimal Value of the Recourse Model

Consider the two-stage recourse model SLPwR with relatively complete fixed recourse. Moreover, assume that the recourse is sufficiently expensive and that $\mathbb{E}_\omega[|\omega|] < \infty$. Under these assumptions the expected recourse function Q is finite by Theorem 3.3.1, so that the model SLPwR has a finite optimal value if it is feasible. For this case, we discuss bounds on the optimal value of SLPwR.

These bounds can be computed before the model SLPwR is actually solved. This means that we can determine *a priori* whether it might be useful to solve SLPwR. Moreover, the differences between the optimal value of SLPwR and its bound have natural interpretations, as we explain below.

The most convenient representation of SLPwR to discuss the bounds is the compact notation

$$\min_{x \in X_0} \mathbb{E}_\omega [g(\omega, x)]$$

where $X_0 := \{x \in \mathbb{R}^n : x \in X,\ Ax = b\}$, and

$$g(\omega, x) := cx + \min_{y \in Y} \{qy : Wy \sim h(\omega) - T(\omega)x\}, \qquad \omega \in \Omega,\ x \in X_0.$$

Let us call its optimal value *TS* (two-stage), and its optimal solution $x^\star \in X_0$:

$$TS := \min_{x \in X_0} \mathbb{E}_\omega [g(\omega, x)] = \mathbb{E}_\omega \left[g(\omega, x^\star)\right].$$

Expected Value Problem and Value of Stochastic Solution

A related problem is the *expected value problem*; it arises when the random vector ω is replaced by its mean value $E[\omega]$. That is, when $(T(\omega), h(\omega))$ are replaced by their means $(\bar{T}, \bar{h})$. Let us call its optimal value EV, and its optimal solution $\bar{x} \in X_0$:

$$EV := \min_{x \in X_0} g(E[\omega], x) = g(E[\omega], \bar{x}).$$

In more detail, the expected value problem is the deterministic linear program

$$\min_{x,y} \left\{ cx + qy : \begin{array}{rr} Ax & = b \\ \bar{T}x + & Wy \sim \bar{h} \\ x \in X & y \in Y \end{array} \right\}.$$

It inherits the algebraic structure of SLPwR, but it is much simpler to solve due to the 'deterministic' modeling of uncertainty with respect to ω. How good is the result of such simplistic modeling? To judge this, we define the so-called *expected result* of the expected value solution *EEV*,

$$EEV := \mathbb{E}_\omega\left[g(\omega, \bar{x})\right].$$

That is, we evaluate the *expected* costs of $\bar{x}$ in the original *stochastic* recourse model. Since $\bar{x}$ is feasible but not necessarily optimal for this recourse model, it follows that

$$TS \leq EEV.$$

The difference is called *Value of Stochastic Solution*, $VSS := EEV - TS$. It indicates the average price one pays for using the simplistic deterministic model rather than the stochastic model.

Remark 3.3.17 If one is willing to replace ω by its mean value, it is also possible to deal with the second-stage constraints as *hard* constraints, by skipping all recourse possibilities. In that case, one finds

$$EV^{HARD} := \min_{x \in X} \left\{ cx : \begin{array}{c} Ax = b \\ \bar{T}x \sim \bar{h} \end{array} \right\}.$$

Exercise 3.3.18 Prove that $EV \leq EV^{HARD}$. ◁

Wait-and-See Problem and Expected Value of Perfect Information

Let us now consider the wait-and-see problem, corresponding to SLPwR. That is, let us assume that it is possible to wait for the actual value of ω before deciding upon x (and y). For each value of ω we may have a different optimal solution then. Denote this optimal solution by $\hat{x}(\omega)$. Denoting the mean value of the wait-and-see problem by *WS*, we have

$$WS := \mathbb{E}_\omega\left[\min_{x \in X_0} g(\omega, x)\right] = \mathbb{E}_\omega\left[g(\omega, \hat{x}(\omega))\right].$$

Since the optimal solution $x^\star$ to the original recourse model is feasible for each $\omega \in \Omega$, but not necessarily optimal for each $\omega \in \Omega$, it holds that

$$WS \leq TS.$$

The difference is called the *Expected Value of Perfect Information*, $EVPI := TS - WS$. It indicates the average price one pays for knowing only the distribution of the random parameters rather than their actual values.

We conclude that under very weak conditions the wait-and-see model provides a lower bound for the optimal value of the recourse model SLPwR, whereas the expected result gives an upper bound.

The question comes up: is it possible to relate the optimal value *EV* of the expected value problem itself (rather than its expected result *EEV*) with *TS*? The following results appear to be true:

(a) If the recourse structure (Y, q, W) is fixed, then $EV \leq TS$
(b) If in addition T is fixed (only $h(\omega)$ is random) then $EV \leq WS$

Exercise 3.3.19 Prove this. Hint: Use Jensen's inequality. In case (a), it is true that for each ω the function $g(\omega, \cdot)$ is convex, whereas in the second case g is convex jointly in (ω, x). ◁

3.3.5 *Approximation of Distributions*

In many situations, discrete distributions are more easy to deal with in algorithms than continuous distributions. Moreover, the expected value of a continuous function of a random vector can be approximated arbitrarily close by approximating the distribution by a discrete one (sampling). However, the number of realizations needed to find a close approximation may be very large. Therefore, in the case of recourse models we are interested in *small-sized* discrete approximations of the given distribution of the random vector ω. Moreover, it would be nice to calculate (using such approximations only) lower and upper bounds on the expected recourse function. This idea leads to nice theory, used by e.g. Kall and Stoyan [13] and Frauendorfer [7], in algorithms to solve recourse models.

Jensen and Edmundson-Madansky Inequalities in One Dimension

Let ω be a random variable with bounded support $\Omega = [a, b]$ and expected value $\bar{\omega} \in (a, b)$, and cdf F_0. Let φ be a convex function on Ω. Since any tangent line minorizes a convex function, and any chord gives larger function values, it holds that

$$\varphi(\bar{\omega}) + u \cdot (t - \bar{\omega}) \leq \varphi(t) \leq \frac{b-t}{b-a}\varphi(a) + \frac{t-a}{b-a}\varphi(b), \quad t \in \Omega,$$

where u is an element of the subgradient of φ at $\bar{\omega}$, it follows by taking expectations that

$$\varphi(\bar{\omega}) \leq \mathbb{E}_\omega [\varphi(\omega)] \leq \frac{b-\bar{\omega}}{b-a}\varphi(a) + \frac{\bar{\omega}-a}{b-a}\varphi(b).$$

That is,

$$\int \varphi(t)\, dF_*(t) \leq \int \varphi(t)\, dF_0(t) \leq \int \varphi(t)\, dF^*(t)$$

where the cdf F_* represents the one-point distribution in $\bar{\omega}$, and the cdf F^* represents the two-point distribution having probability $\frac{b-\bar{\omega}}{b-a}$ at $\omega = a$ and $\frac{\bar{\omega}-a}{b-a}$ at $\omega = b$. Note that both F_* and F^* have mean value $\bar{\omega}$, just as F_0 itself. The first inequality is known as *Jensen's inequality*, and the second as an *Edmundson-Madansky inequality*.

Therefore, denoting the family of all distributions on Ω with mean value $\bar{\omega}$ by $\mathcal{F}$, we get

$$\min_{F \in \mathcal{F}} \int \varphi(t)\, dF(t) = \int \varphi(t)\, dF_*(t),$$

$$\max_{F \in \mathcal{F}} \int \varphi(t)\, dF(t) = \int \varphi(t)\, dF^*(t).$$

The Jensen and Edmundson-Madansky inequalities have nice properties:

(i) They can be calculated easily, since only one or two-point distributions are involved.
(ii) They provide a lower bound and an upper bound, that is valid (and sharp in $\mathcal{F}$) uniformly for all *convex* functions φ.

Both properties are useful in stochastic programming; the second one, because in SLPwR we have $\varphi(\omega) = v(x, \omega)$ depending on the decision x as well. (Recall that the second-stage optimal value is a convex function of ω if we have fixed q and W, that is, if only $h(\omega)$ and $T(\omega)$ are random.)

The obvious disadvantage of these bounds is that they are not close approximations of $\int \varphi(t)\, dF_0(t)$. However, by *partitioning* of Ω, and using more detailed information of the distribution F_0, that is, by conditioning, it is possible to improve the bounds. This can be seen as follows.

Partition $\Omega = [a, b]$ into $\bigcup_{k=1}^{K} \Omega_k$, where $\Omega_k := (s_{k-1}, s_k]$, $k = 1, \dots, K$, and $a = s_0 < s_1 < \cdots < s_{K-1} < s_K = b$, such that $p_k := \Pr\{\omega \in \Omega_k\} = F_0(s_k) - F_0(s_{k-1}) > 0$. If we denote the conditional cdf of ω given $\omega \in \Omega_k$ by F_k, then

$$F_k(t) = \begin{cases} 0, & t \leq s_{k-1}, \\ \dfrac{F_0(t) - F_0(s_{k-1})}{F_0(s_k) - F_0(s_{k-1})}, & s_{k-1} < t \leq s_k, \\ 1, & t > s_k. \end{cases}$$

Obviously, we have $F_0(t) = \sum_{k=1}^{K} p_k F_k(t)$, $t \in \mathbb{R}$, so that

$$\int \varphi(t)\, dF_0(t) = \sum_{k=1}^{K} p_k \int \varphi(t)\, dF_k(t).$$

By applying the Jensen and Edmundson-Madansky inequalities to each subinterval separately, we get the following result.

Theorem 3.3.20 *For any convex function φ on $\Omega = [a, b]$ it holds that*

$$\int \varphi(t)\, d\underline{F}(t) \leq \int \varphi(t)\, dF_0(t) \leq \int \varphi(t)\, d\overline{F}(t),$$

where $\underline{F}$ is the cdf of the K-point distribution, that gives probability p_k to $\mu_k = \mathbb{E}_\omega[\omega \mid \omega \in \Omega_k] = \int t\, dF_k(t)$, $k = 1, \ldots, K$, and $\overline{F}$ is the cdf of the $(K+1)$-point distribution, that gives probability $p_{k+1}(1 - q_{k+1}) + p_k q_k$ to s_k, $k = 0, \ldots, K$, where $q_k := \dfrac{\mu_k - s_{k-1}}{s_k - s_{k-1}}$, $k = 1, \ldots, K$ (and $p_0 q_0 := 0$, $p_{K+1}(1 - q_{K+1}) := 0$).

Exercise 3.3.21

(a) Prove Theorem 3.3.20.
(b) Show that these generalized Jensen and Edmundson-Madansky inequalities can be interpreted as piecewise linear approximations of φ.
(c) Show that by adding a new point into the partition, the corresponding lower bound cannot decrease, and the corresponding upper bound cannot increase.

◁

From this introduction it will be clear that it is possible to generate an algorithm, that at each iteration refines the current partition by adding a new point s. By choosing s in the subinterval where the local Edmundson-Madansky upper bound differs most from the local Jensen lower bound, it is to be expected that the most improvement will be obtained.

Bounds of this type are useful too in case the distribution of ω is not completely specified, e.g. if only the mean and the range are known.

The results presented above can be generalized to higher dimensions. Another generalization applies to convex-concave saddle functions. See [13] and [7].

3.4 Algorithms for Recourse Models

From a theoretical point of view, the recourse model SLPwR is a *nice* nonlinear programming problem, that is *difficult to solve*, except in *special cases*:

nice: it can be shown that SLPwR has linear constraints and a convex objective function $cx + Q(x)$. So there is a rather complete optimization theory, with necessary and sufficient conditions of optimality (Karush–Kuhn–Tucker) and many good algorithms.

difficult to solve: in general it is very demanding to generate function evaluations (for $Q(x)$ or for its gradient $\nabla Q(x)$). Indeed, in order to calculate $Q(x)$ for a fixed value x_0 one has, in principle, to solve the second-stage LP for each $\omega \in \Omega$, and on top of that one has to calculate the expected value. Basically, the last operation is to be classified as calculating a multiple integral.

special cases: the difficulties above disappear when the recourse structure is simple, as will be discussed in detail later. Another special case for which the problem becomes easier is when Ω is finite, since then SLPwR is equivalent to a large scale linear programming problem (see Sect. 3.2.1).

In Appendix E we sketch some main ideas behind general algorithms for convex non-linear optimization problems. Some of these ideas will be worked out in more detail in the context of specific algorithms for stochastic linear programming models.

3.4.1 Simple Recourse

We consider algorithms for simple recourse models with only the right-hand side $h(\omega)$ random, that is, $T(\omega) = T$ is fixed. We conveniently assume $h(\omega) = \omega \in \mathbb{R}^m$. As discussed in Sect. 3.3.2, in this case the expected value function Q is separable in the tender variables $z := Tx$, so that

$$Q(x) = \sum_{i=1}^{m} Q_i(z_i) \quad \text{where}$$

$$Q_i(z_i) = q_i^+ \mathbb{E}_{\omega_i}\left[(\omega_i - z_i)^+\right] + q_i^- \mathbb{E}_{\omega_i}\left[(\omega_i - z_i)^-\right], \quad z_i \in \mathbb{R},$$

at least if $q_i^+ + q_i^- \geq 0$, $i = 1, \ldots, m$. This allows us to present algorithms for simple recourse models in a one-dimensional setting. To keep the notation simple, we use $Q : \mathbb{R} \mapsto \mathbb{R}$ below, and explicitly mention when we use Q to represent the n-dimensional expected value function. Similarly, ω denotes a random variable unless specified otherwise.

Discrete Distributions

First consider the case that ω follows a finite discrete distribution, say $\omega \in \Omega = \{\omega^1, \ldots, \omega^S\}$ with $\omega^1 < \omega^2 < \cdots < \omega^S$, $\Pr\{\omega = \omega^s\} = p^s$ and $\mathbb{E}_\omega[\omega] = \mu$.

From Theorem 3.3.1 we know that Q is piecewise linear, with right derivative

$$Q'_+(x) = -q^+ + \left(q^+ + q^-\right) F(x), \quad x \in \mathbb{R},$$

where F is the (piecewise constant) cdf of ω. Thus, for $x \in [\omega^s, \omega^{s+1})$ (with $\omega^0 = -\infty$, $\omega^{S+1} = +\infty$), we have

$$Q'_+(x) = -q^+ + \left(q^+ + q^-\right) F(\omega^s) := \delta^s,$$

with $-q^+ = \delta^0 < \delta^1 < \cdots < \delta^S = q^-$. It is possible to represent this piecewise linear function in an LP setting. Corresponding to each interval $[\omega^s, \omega^{s+1}]$, $s = 0, \ldots, S$, we define an additional variable u^s with appropriate lower and upper bounds (see below). Then, for $x \in \mathbb{R}$,

$$\begin{aligned} Q(x) = \min_u \ & \textstyle\sum_{s=0}^{S} \delta^s u^s \\ \text{s.t.} \ & \sum_{s=0}^{S} u^s = x - \mu \\ & u^0 \leq \omega^1 - \mu \\ & 0 \leq u^s \leq \omega^{s+1} - \omega^s, \ s = 1, \ldots, S \end{aligned}$$

Hence, the simple recourse model with discretely distributed right-hand side $\omega = (\omega_1, \ldots, \omega_m)$, $\Omega_i = \{\omega_i^1, \ldots, \omega_i^{S_i}\}$, $\omega_i^1 < \omega_i^2 < \cdots < \omega_i^{S_i}$, $\Pr\{\omega_i = \omega_i^s\} = p_i^s$ and $\mathbb{E}_{\omega_i}[\omega_i] = \mu_i$, is equivalent to the LP problem

$$\begin{aligned} \min_{\substack{x \in \mathbb{R}^n_+ \\ u}} \ & cx + \sum_{i=1}^{m} \sum_{s=0}^{S_i} \delta_i^s u_i^s \\ \text{s.t.} \ & Ax = b \\ & \sum_{s=0}^{S_i} u_i^s - T_i x = \mu_i \quad i = 1, \ldots, m \\ & u_i^0 \leq \omega_i^1 - \mu_i \qquad i = 1, \ldots, m \\ & 0 \leq u_i^s \leq \omega_i^{s+1} - \omega_i^s \ s = 1, \ldots, S_i, \ i = 1, \ldots, m \end{aligned} \tag{3.24}$$

The proof of this result is left as exercise (see Assignment R3).

Actually, we already showed in Sect. 3.2 that any two-stage recourse problem with discrete random parameters is equivalent to a large-scale LP problem. However, using the naive approach of Sect. 3.2 results in an LP with $n + 2mS$ variables and $m_1 + mS$ constraints, where $S = \prod_{i=1}^m S_i$ (assuming independence of the components of ω). For realistic numbers of random variables and realizations, this is a huge LP problem. On the other hand, it is readily verified that the LP problem (3.24) has only $n + m + \sum_{i=1}^m S_i$ variables and $m_1 + 2m + \sum_{i=1}^m S_i$ constraints (which are mainly simple bounds). This is only marginally larger than the underlying LP problem (in which all random parameters are replaced by their expected values), which has n variables and $m_1 + m$ constraints. Typically, the solution time for (3.24) is only three times as large as that for the underlying LP problem.

We conclude that simple recourse problems with discretely distributed right-hand side parameters (and fixed technology matrix T) can be solved very efficiently.

Continuous Distributions

Assume that the random variable ω is continuously distributed, with mean value $\mathbb{E}_\omega[\omega] = \mu$ and bounded support $\Omega = [a, b]$ (if necessary, truncate the support). A general scheme to solve this simple recourse problem could be

(1) Approximate the distribution of ω by a suitable discrete distribution.
(2) Solve the resulting simple recourse problem *with discretely distributed right-hand side*, that is, solve the corresponding LP problem (3.24).
(3) Stop if the current solution is satisfactory, otherwise return to (1) and update the approximating distribution.

The general theory of Sect. 3.3.5 teaches that, since the one-dimensional value function $v(\omega - x) = q^+(\omega - x)^+ + (\omega - x)^-$ is convex in ω, a uniform lower bound for Q is obtained if a Jensen type distribution is used, whereas a uniform upper bound can be obtained using an Edmundson-Madansky type approximate distribution. It is clear that the performance of an actual algorithm is to a large extent determined by the construction of the approximating distributions, that is, by the updating strategy of the partition of Ω. Obviously, the difference between the objective value of the current solution in both approximations can then be used as a stopping criterion. Below we will show that a more efficient approach is possible in the special case at hand. In particular, we will see that there is an obvious strategy to partition, which makes the explicit construction of an upper bound superfluous.

The crux of the approach presented here, is the following property of the one-dimensional function Q (see Theorem 3.3.15): outside the support $\Omega = [a, b]$ the function Q coincides with its asymptotes. Equivalently, this can be formulated as

$$Q(x) = v(\mu - x), \qquad x \notin (a, b). \tag{3.25}$$

In the first iteration of the prospective algorithm, our partition of the support is simply $\Omega = \Omega$, and the Jensen type approximating distribution is given by $\Pr\{\xi = \mu\} = 1$. Let x^1 denote an optimal solution of the corresponding LP problem, and let $\mathcal{Q}^1(x^1)$ be the objective value (without loss of generality, we assume here that $c = 0$). If $x^1 \notin (a, b)$, it follows from (3.25) that $\mathcal{Q}^1(x^1) = Q(x^1)$, implying that x^1 is an optimal solution of the true simple recourse problem, and the algorithm stops. Otherwise, we proceed with iteration 2.

First, partition $\Omega = \Omega_1 \cup \Omega_2$, where the solution x^1 is used to define the subsets: $\Omega_1 := (a, x^1]$ and $\Omega_2 := (x^1, b]$. The approximating distribution is $\Pr\{\xi = \mu_k\} = \Pr\{\omega \in \Omega_k\} =: p_k$, with $\mu_k := \mathbb{E}_\omega[\omega \mid \omega \in \Omega_k]$, $k = 1, 2$. Let x^2 and $\mathcal{Q}^2(x^2)$ be an optimal solution and the optimal value of the corresponding LP problem, respectively. Next, we observe that for $x \in \mathbb{R}$,

$$Q(x) = p_1 Q_1(x) + p_2 Q_2(x) \tag{3.26}$$

where $Q_k(x) := \mathbb{E}_\omega [v(\omega - x) \mid \omega \in \Omega_k]$, $k = 1, 2$. (Verify this!) By construction, the optimal solution x^1 found in the first iteration satisfies $x^1 \notin (a, x^1)$ and $x^1 \notin (x^1, b)$. Therefore, combining (3.25) and (3.26), we obtain

$$Q(x^1) = p_1 v(\mu_1 - x^1) + p_2 v(\mu_2 - x^1) = \mathbb{E}_\xi \left[v(\xi - x^1) \right] = \mathcal{Q}^2(x^1).$$

That is, the approximation $\mathcal{Q}^2$ of Q is *exact* at the solution x^1 found in the first iteration! Consequently, there is no need to construct an upper bound (e.g. of the Edmundson-Madansky type) to get an error estimate: the estimate $\mathcal{Q}^2(x^1) - \mathcal{Q}^1(x^1) = Q(x^1) - \mathcal{Q}^1(x^1)$ is available at no extra costs.

Thus, in iteration $t + 1$ of the algorithm we obtain x^{t+1} and $\mathcal{Q}^{t+1}(x^{t+1})$, and also $Q(x^t)$ where x^t is the solution found in iteration t. The algorithm stops with solution x^t if $Q(x^t) - \mathcal{Q}^t(x^t) \leq \varepsilon^*$, where ε^* is some given accuracy parameter. If this stopping criterion is not satisfied, the partition of Ω is refined as follows: identify $\Omega_k = (s_{k-1}, s_k]$ such that $x^{t+1} \in \Omega_k$, and set

$$\Omega_k \to (s_{k-1}, x^{t+1}] \cup (x^{t+1}, s_k],$$

and proceed with the next iteration.

Example 3.4.1 Let a one-dimensional simple recourse problem be specified with right-hand side ω uniformly distributed on $(0, 4)$, $q^+ = 1$, $q^- = 3$, and $c = 0$, see Fig. 3.3.

- *Iteration 1*. The partition of Ω is simply Ω itself, so that the discrete Jensen approximate distribution is given by $\Pr\{\xi_1 = \mu\} = 1$, with $\mu := \mathbb{E}_\omega[\omega] = 2$. Solving the corresponding approximating simple recourse problem (3.24) gives $x^1 = 2$ (with $\mathcal{Q}^1(2) = 0$).
- *Iteration 2*. Partition $\Omega = (0, 2] \cup (2, 4]$ with conditional mean values 1 and 3, respectively. The approximate discrete distribution is $\Pr\{\xi_2 = 1\} = \Pr\{\xi_2 = 3\} = 1/2$, leading to a solution $x^2 = 1$. Moreover, we obtain the exact value of $Q(x^1) = 1/2 v(1 - x^1) + 1/2 v(3 - x^1)$.
- *Iteration 3*. Since $x^2 \in (0, 2]$, refine the partition of Ω to become $\Omega = (0, 1] \cup (1, 2] \cup (2, 4]$, and solve the problem with right-hand side ξ_3, with $\Pr\{\xi_3 = 1/2\} = \Pr\{\xi_3 = 3/2\} = 1/4$ and $\Pr\{\xi_3 = 3\} = 1/2$. A solution is $x^3 = 1/2$, and we obtain $Q(x^2) = \mathbb{E}_{\xi_3}\left[v(\xi_3 - x^2)\right]$.
- *Iteration 4*. Set $\Omega = (0, 1/2] \cup (1/2, 1] \cup (1, 2] \cup (2, 4]$, leading to a discrete distribution with $\Pr\{\xi_4 = 1/4\} = \Pr\{\xi_4 = 3/4\} = 1/8$, $\Pr\{\xi_4 = 3/2\} = 1/4$ and $\Pr\{\xi_4 = 3\} = 1/2$. A solution to the corresponding problem is $x^4 = 3/4$. Using $Q(x^3) = \mathbb{E}_{\xi_4}\left[v(\xi_4 - x^3)\right]$, we find that $Q(x^3) - \mathcal{Q}^3(x^3) \leq 0.01$.

Actually, $x^2 = 1$ is the optimal solution of this problem. By construction, x^2 is an optimal solution in all iterations $t \geq 2$ and its optimal value is known. However, as illustrated here, this may not be recognized by the algorithm. (Why?) ◇

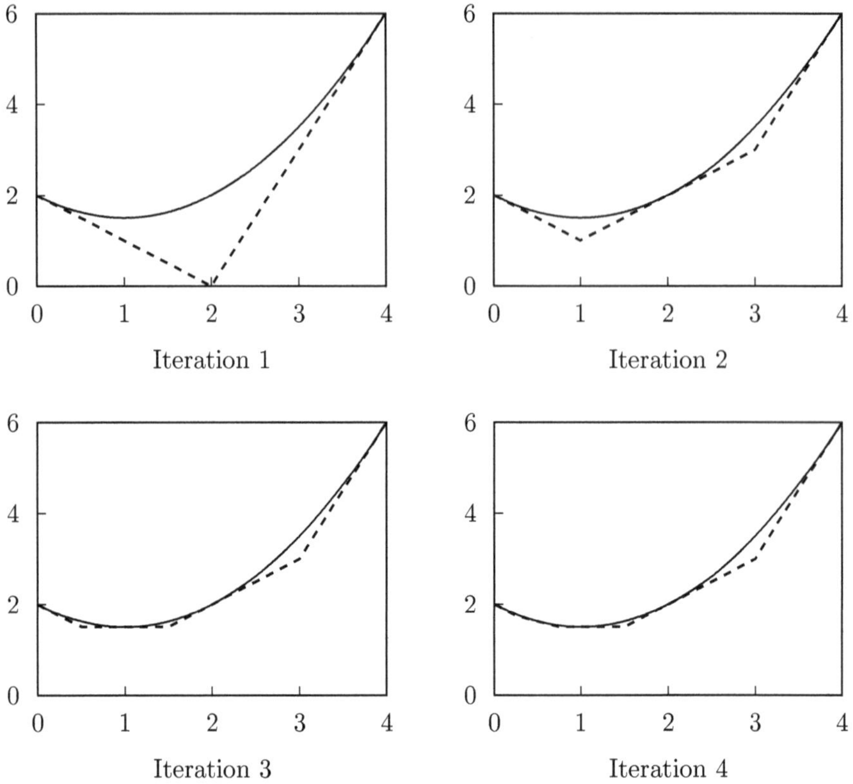

Fig. 3.3 Illustration of the algorithm for simple recourse with continuously distributed right-hand side

Simple Recourse with Random Technology Matrix

For the general simple recourse model with random technology matrix $T(\omega)$ (and right-hand side $h(\omega)$) no specific algorithms are available; thus, observing that simple recourse is complete, such SR problems can be solved by complete recourse algorithms as discussed in the next section. However, as a spin-off of results obtained for integrated chance constraints (see Chap. 6), Klein Haneveld and Van der Vlerk [22] proposed an algorithm for simple recourse models with *discretely distributed technology matrix* and right-hand side vector.

Once more we consider the structure of the SR expected value function Q. Let ω_i denote the subvector of components of ω on which $T_i(\omega)$ and $h_i(\omega)$ actually depend (note that ω_i and ω_j, $i \neq j$, may overlap). Then $(T_i(\omega), h_i(\omega)) = (T_i(\omega_i), h_i(\omega_i))$, $i = 1, \ldots, m$, so that the function Q can be written in closed form:

$$Q(x) = \sum_{i=1}^{m} Q_i(x), \qquad x \in \mathbb{R}^n, \tag{3.27}$$

with, for $i = 1, \dots, m$,

$$\begin{aligned} Q_i(x) &:= q_i^+ \mathbb{E}_{\omega_i}\left[\left(h_i(\omega_i) - T_i(\omega_i)x\right)^+\right] + q_i^- \mathbb{E}_{\omega_i}\left[\left(h_i(\omega_i) - T_i(\omega_i)x\right)^-\right] \\ &= q_i^-\left(\bar{T}_i x - \bar{h}_i\right) + \left(q_i^+ + q_i^-\right)\mathbb{E}_{\omega_i}\left[\eta_i(x, \omega_i)^-\right], \end{aligned}$$

utilizing the well-known closed form for the underlying separable function v (see e.g. Theorem 3.3.15), and the notation $\eta_i(x, \omega_i) := T_i(\omega_i)x - h_i(\omega_i)$ and $\left(\bar{T}_i, \bar{h}_i\right) := \mathbb{E}_{\omega_i}\left[\left(T_i(\omega_i), h_i(\omega_i)\right)\right]$. Note that for each Q_i the expectation is taken with respect to the marginal distribution of ω_i.

Assume that ω, and therefore each ω_i, is discretely distributed, with given probabilities $\Pr\left\{\omega_i = \omega_i^s\right\} = p_i^s$, $s \in S_i$. Then, as shown in the proof of Theorem 6.2.11, it holds

$$\mathbb{E}_{\omega_i}\left[\eta(x, \omega_i)^-\right] = \max_{K \subset S_i} \sum_{k \in K} -p_i^k \eta(x, \omega_i^k),$$

so that, defining $K_i^t := \left\{s \in S_i : \eta_i(x^t, \omega_i^s)^- > 0\right\}$ for any fixed x^t, we find that

$$\begin{aligned} L_i^t(x) &:= q_i^-\left(\bar{T}_i x - \bar{h}_i\right) + \left(q_i^+ + q_i^-\right) \sum_{k \in K_i^t} -p_i^k \eta_i(x, \omega_i^k) \\ &= q_i^-\left(\bar{T}_i x - \bar{h}_i\right) + \left(q_i^+ + q_i^-\right) \sum_{k \in K_i^t} p_i^k \left(h_i^k - T_i^k x\right) \end{aligned}$$

is a linear lower bound for Q_i, which is sharp at x^t. Consequently,

$$L^t(x) := \sum_{i=1}^m L_i^t(x), \qquad x \in \mathbb{R}^n,$$

is a linear lower bound for the expected value function Q, which is sharp at x^t. That is, for arbitrary fixed x^t, $L^t(x)$ is a linear *optimality cut* for Q at x^t as needed in (variants of) the L-shaped algorithm discussed in the next section.

An algorithm using these tailor-made optimality cuts can be expected to perform relatively well, since each evaluation of Q (which also yields the corresponding optimality cut) using (3.27) requires only $\sum_{i=1}^m S_i$ evaluations of v, instead of (at most) $\prod_{i=1}^m S_i$ evaluations as needed in L-shaped algorithms for general complete recourse models.

3.4.2 *Fixed Recourse*

A more general case for which more or less efficient algorithms exist, is the (complete) recourse model with discretely distributed parameters $T(\omega)$ and $h(\omega)$, that is, for the case with $\omega \in \Omega = \{\omega^1, \ldots, \omega^S\}$ with respective probabilities p_s. Typically, such algorithms take as starting point the equivalent large-scale LP problem, which can be written (in canonical form) as

$$\begin{aligned} \min_{x, y^s} \; & cx + \sum_{s=1}^{S} p_s q y^s \\ \text{s.t.} \; & Ax = b \\ & W y^s + T^s x = h^s, \; s = 1, \ldots, S \\ & x \in \mathbb{R}^n_+, \; y^s \in \mathbb{R}^p_+, \; s = 1, \ldots, S \end{aligned} \tag{3.28}$$

where $T^s := T(\omega^s)$ and $h^s := h(\omega^s)$.

From a computational point of view, the first-stage variables x are complicating in (3.28), since they destroy the separability of the problem. Indeed, if x is given, the computation of the optimal second-stage solutions y^s (i.e., of $Q(x)$) separates in S second-stage problems

$$v^s(x) := \min_{y^s} \left\{ q y^s : W y^s = h^s - T^s x, \; y^s \in \mathbb{R}^p_+ \right\}.$$

This observation on the *structure* of the problem underlies a number of algorithms. Below, we outline a few important examples of them.

The L-Shaped Algorithm

First we assume that the recourse structure is complete (and sufficiently expensive). Then from Theorem 3.3.1 we know that the expected value function Q is convex polyhedral. Moreover, for any $\bar{x} \in \mathbb{R}^n$ it holds

$$v^s(\bar{x}) = \max_{\lambda} \left\{ \lambda(h^s - T^s \bar{x}) : \lambda W \leq q \right\},$$

and any optimal solution λ_s^* of this dual second-stage problem is a subgradient of v^s at $\bar{x}$. As argued in Sect. 3.3.1, it follows that

$$u := -\sum_{s=1}^{S} p_s \lambda_s^* T^s \in \partial Q(\bar{x})$$

so that

$$Q(x) \geq Q(\bar{x}) + \langle u, x - \bar{x} \rangle \quad \forall x \in \mathbb{R}^n.$$

In other words, $Q(\bar{x}) + \langle u, x - \bar{x} \rangle$ is a supporting hyperplane for Q at $\bar{x}$. This insight can be used to iteratively construct increasingly better approximations of the function Q, as implemented in the L-shaped method of Van Slyke and Wets [49]. Next we present the basic scheme of this algorithm, of which many variants are known in the literature (see e.g. Mayer [27]).

The algorithm starts with solving the LP problem

$$\begin{aligned} \min_x \ & cx \\ \text{s.t.} \ & Ax = b \\ & x \in \mathbb{R}^n_+ \end{aligned}$$

that is, the expected recourse costs Q are ignored. Let x^1 be an optimal solution. For $s = 1, \ldots, S$, the dual formulation of the second-stage problem

$$v^s(x^1) = \max_\lambda \left\{ \lambda(h^s - T^s x^1) : \lambda W \leq q \right\}$$

is solved, to obtain optimal solutions λ_1^s, $s = 1, \ldots, S$. Next, we compute

$$\begin{aligned} Q(x^1) &= \sum_{s=1}^{S} p_s \lambda_1^s (h^s - T^s x^1) \\ u^1 &= -\sum_{s=1}^{S} p_s \lambda_1^s T^s, \end{aligned}$$

and use this to define the *optimality cut*

$$\theta \geq Q(x^1) + \langle u^1, x - x^1 \rangle =: \alpha^1 + \beta^1 x, \quad x \in \mathbb{R}^n,$$

where the variable θ is introduced to replace the function $Q(x)$ in subsequent computations (see below).

In the second iteration, the *current problem* to be solved is the LP problem

$$\begin{aligned} \min_{x,\theta} \ & cx + \theta \\ \text{s.t.} \ & Ax = b \\ & \theta \geq \alpha^1 + \beta^1 x \\ & x \in \mathbb{R}^n_+, \quad \theta \in \mathbb{R}. \end{aligned}$$

We see that the polyhedral convex function Q, which by definition is equal to the pointwise maximum of all its finitely many supporting hyperplanes, i.e.

$$Q(x) = \max_{k=1,\dots,K} \{d_k + e_k x\}, \qquad x \in \mathbb{R}^n,$$

is approximated by the variable θ, which is bounded from below by only one of these supporting hyperplanes (called optimality cuts). The idea of the algorithm is to add optimality cuts to the current problem one by one, thus constructing a *sufficiently good description of Q at the optimal solution*. Hopefully, this occurs when only a fraction of all possible cuts are generated.

In iteration t, let the current problem be

$$\begin{aligned} \min_{x,\theta}\ & cx + \theta \\ \text{s.t.}\ \ & Ax = b \\ & \theta \geq \alpha^j + \beta^j x, \quad j = 1, \dots, t-1 \\ & x \in \mathbb{R}^n_+, \quad \theta \in \mathbb{R}, \end{aligned}$$

and let (x^t, θ^t) be an optimal solution. $Q(x^t)$ is computed as indicated above, which also provides the information needed to construct the optimality cut $\theta \geq \alpha^t + \beta^t x$ which is sharp at x^t. With this cut added to the set of optimality cuts, proceed with iteration $t+1$, unless x^t is an optimal solution of the recourse problem.

Exercise 3.4.2

(a) Show that if $\theta^t \geq Q(x^t)$, then x^t is an optimal solution of the recourse problem.
(b) Show that the L-shaped algorithm finds an optimal solution in finitely many iterations (assuming that there is one).

◁

Remark 3.4.3 The same algorithm applies to the deterministic generic mixed-integer problem

$$\begin{aligned} \min_{x,y}\ & cx + qy \\ \text{s.t.}\ \ & Ax = b \\ & Wy + Tx = h \\ & x \in \mathbb{Z}^n_+,\ y \in \mathbb{R}^p_+. \end{aligned}$$

Here, the integer variables x are complicating, since for fixed x the remaining problem

$$\min_y \{qy : Wy = h - Tx,\ y \in \mathbb{R}^p_+\} =: LP(x)$$

is an LP (and the value function $LP(x)$ is convex in x). In the setting of MIP problems, the algorithm is known as *Benders' decomposition* [2] .

Example 3.4.4 Consider the (simple) recourse problem

$$\min_{x \in [0,5]} x/2 + Q(x),$$

with $Q(x) = \mathbb{E}_\omega\left[(\omega - x)^+\right] + 3\mathbb{E}_\omega\left[(\omega - x)^-\right]$, $x \in \mathbb{R}$, and ω following a uniform discrete distribution on $\{1, 2, 3, 4\}$. Figure 3.4 illustrates its solution by the L-shaped algorithm. ◇

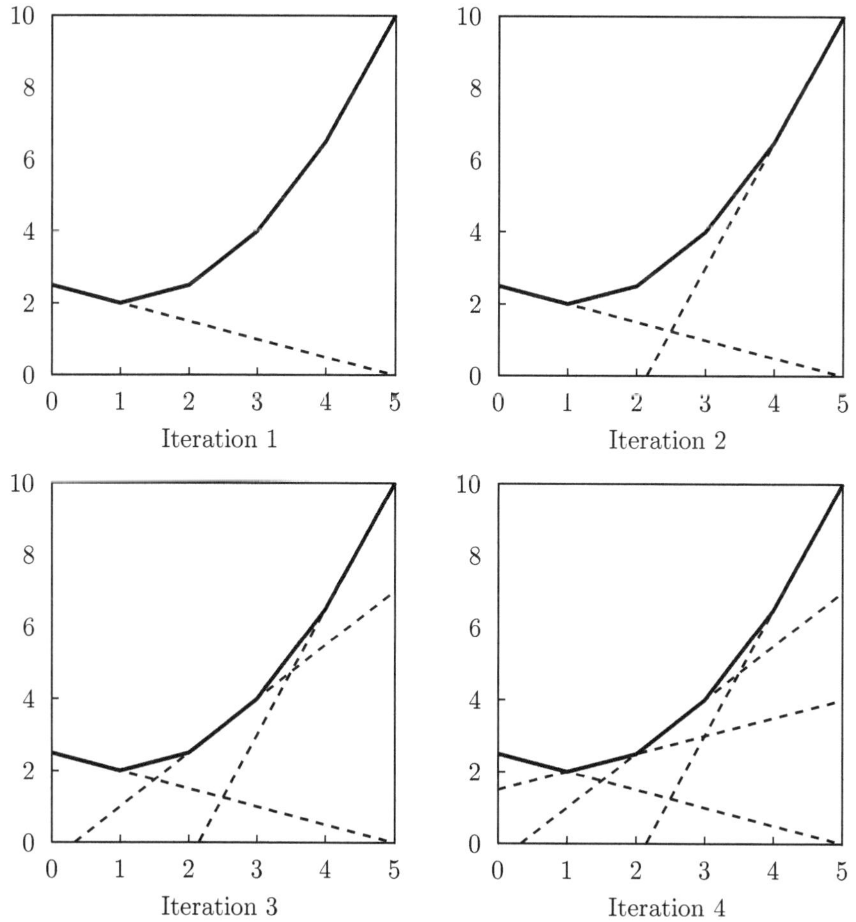

Fig. 3.4 Illustration of the L-shaped algorithm

Recall from Sect. 3.2.3 that if the recourse is not relatively complete, the restricted recourse possibilities give rise to additional constraints on the first-stage decision variables x:

$$\Pr\{\omega \in \Omega : v(h(\omega) - T(\omega)x) = +\infty\} = 0. \tag{3.29}$$

Since the distribution of ω is discrete here, the set of first-stage decisions satisfying (3.29) is polyhedral, and can therefore be described as the intersection of finitely many half-spaces. The L-shaped algorithm iteratively generates the corresponding separating hyperplanes, called *feasibility cuts*, similar to the optimality cuts presented above.

Let (x^t, θ^t) be an optimal solution of the current problem in iteration t. For $s = 1, \ldots, S$, solve the (Phase-I Simplex) problems

$$\begin{aligned} w^s(x^t) &:= \min_{y^s} \left\{ e(y^+ + y^-) : Wy^s + y^+ - y^- = h^s - T^s x^t,\ y^s \in \mathbb{R}^p_+ \right\} \\ &= \max_{\nu} \left\{ \nu(h^s - T^s x^t) : \nu W \le 0,\ |\nu| \le e,\ \nu \in \mathbb{R}^m \right\}, \end{aligned}$$

where $e = (1, \ldots, 1) \in \mathbb{R}^m$, and let ν_t^s be their respective optimal solutions.

Exercise 3.4.5 Show that $v^s(x^t) = +\infty$ if and only if $w^s(x^t) > 0$. ◁

It follows that if $w^{\bar{s}}(x^t) > 0$ for some $\bar{s} \in \{1, \ldots, S\}$, then $Q(x^t) = +\infty$ and x^t is not feasible with respect to the constraint (3.29). The feasibility cut (also known as *induced constraint*)

$$\nu_t^{\bar{s}} \left(h^{\bar{s}} - T^{\bar{s}} x \right) \le 0$$

cuts off x^t but is satisfied by all x such that $Q(x) < +\infty$.

Exercise 3.4.6 Verify this. ◁

With this cut added to the set of feasibility cuts, the algorithm proceeds with iteration $t+1$. If $w^s(x^t) = 0$ for all s, then an optimality cut is generated as described above.

Continuous Distributions

As mentioned in the introduction of this section, if the distribution of the underlying random vector $\omega \in \mathbb{R}^r$ is continuous, it is in general not possible to solve the recourse problem due to the insurmountable problems caused by evaluating the expected value function Q, which amounts to evaluating an r-dimensional integral. In this situation one is forced to resort to solving an approximation of the recourse problem, using discrete approximations of the distribution of ω.

One option is to specify such an approximating distribution, possibly based on Jensen's inequality or the Edmundson-Madansky inequality, and solve the corresponding problem by e.g. (variants of) the L-shaped method. Although it is possible to obtain information on the quality of the approximate solution thus obtained (see Sect. 3.3.4), it may be very difficult to specify 'good' discrete distributions a priori.

Alternatively, suitable distributions giving upper and lower bounds can be constructed iteratively, each time updating the distributions guided by the information gathered so far. Obviously, there are many possible rules that can be used to update the distributions, and it is also clear that this choice has a strong influence on the performance of the solution method. Below we discuss updating rules suggested by Kall and Stoyan [13] that appear to work well in practice.

Discrete Approximation

The Discrete Approximation algorithm uses respectively Jensen and Edmundson-Madansky type distributions to obtain lower bounds and upper bounds for the optimal value of the recourse problem. Given a current partition $\Omega = \bigcup_{k-1}^{K_t} \Omega_k$ at iteration t of the algorithm, the question is how this partition should be updated in order to improve the approximation. Given that computation of the Edmundson-Madansky upper bound in a current solution x^t implies evaluation of the second-stage value function in all 2^r vertices of each subset Ω_k, and that a 'bad' subdivision is carried along in all subsequent iterations, it is worthwhile to spend some (CPU) time on answering this question. The following subproblems can be distinguished.

(1) Which subset Ω_k should be partitioned? A reasonable choice appears to be to pick Ω_κ such that

$$\kappa \in \operatorname*{argmax}_{k=1,\dots,K_t} UB_k - LB_k,$$

where

$$LB_k := p_k v(x^t, \mu_k), \qquad UB_k := \sum_{i=1}^{2^r} p_{ki} v(x^t, s^{ki}),$$

with $p_k = \Pr\{\omega \in \Omega_k\}$, $\mu_k = \mathbb{E}_\omega[\omega \mid \omega \in \Omega_k]$, and the p_{ki} are the probabilities assigned to the vertices s^{ki} of Ω_k by the Edmundson-Madansky approximating distribution. That is, LB_k is the Jensen lower bound for $Q(x^t)$ on Ω_k, and UB_k is the Edmundson-Madansky upper bound on Ω_k. (This information is available from previous computations.) Thus, on the set Ω_κ the current approximation of $Q(x^t)$ may be relatively poor, which suggest to improve the approximation here.

(2) Given the choice of the subset which will be partitioned, the next question is how to partition it. Even if we limit the possibilities by assuming that the set will be split in two new subsets, and that the split is along a coordinate direction, a lot of freedom remains. Kall and Stoyan suggest the following approach. For each coordinate $j = 1, \dots, r$

  (i) Try the subdivision at $\mathbb{E}_\omega \left[\omega_j \mid \omega \in \Omega_\kappa\right]$, resulting in subsets $\Omega^1_{\kappa_j}$ and $\Omega^2_{\kappa_j}$, with conditional probabilities π^1_j and π^2_j, respectively.

  (ii) On both sets compute the local Jensen lower bound $LB^i_{\kappa_j}$ and Edmundson-Madansky upper bound $UB^i_{\kappa_j}$, $i = 1, 2$, and compute

$$\pi^1_j \left(UB^1_{\kappa_j} - LB^1_{\kappa_j}\right) + \pi^2_j \left(UB^2_{\kappa_j} - LB^2_{\kappa_j}\right), \tag{3.30}$$

  which is a measure of the improvement that results if this partition is used.

  (iii) Keep the partition that minimizes (3.30).

This is indeed a lot of work (CPU time) to find a 'good' partition. However, numerical evidence indicates that this effort is well spent.

Chapter 4
Stochastic Mixed-Integer Programming

In this chapter we consider a generalization of the recourse model in Chap. 3, obtained by allowing integrality restrictions on some or all of the decision variables. First we give some motivation why such *mixed-integer recourse models* are useful and interesting. Following the presentation of the general model, we give several examples of applications. Next we discuss mathematical properties of the general model as well as the so-called simple integer recourse model, which is the analogue of the continuous simple recourse model discussed in Sect. 3.3.2. We conclude this chapter with an overview of available algorithms.

4.1 Mixed-Integer Recourse Models

In Chap. 1 we mentioned the extensive possibilities of linear programming as a tool for analyzing decision problems. An even more powerful tool is obtained by allowing integrality restrictions on some of the variables, thus entering the field of mixed-integer programming. There can be several reasons to include integer variables in an LP model:

(i) *Natural integrality of decision variables*: In some applications decision variables represent indivisible entities. Sometimes it is still justified to model this by continuous variables (e.g. the number of newspapers in the 'Newsboy problem'), but in other cases this is incorrect (e.g. if the variable represents the number of oil tankers to be built next year).

(ii) *Yes/no, on/off decisions*: Such decisions are modeled by $0-1$ variables (binary variables). Continuous variables taking values in $[0, 1]$ are useless here, since a fractional solution with values in $(0, 1)$ is meaningless.

W. K. Klein Haneveld et al., *Stochastic Programming*, Graduate Texts in Operations Research, https://doi.org/10.1007/978-3-030-29219-5_4

(iii) *Artificial indicator variables for conditional linear constraints*: Binary variables often have this function in LP formulations of Combinatorial Optimization problems.
(iv) *Satisfy k out of n constraints*: This can also be modeled by binary variables. See also Sect. 5.3 on chance constraints with discrete distributions.

Mixed-integer programming (MIP) models are widely used to analyze a variety of problems, including applications such as production planning, inventory control, logistics, and scheduling. This is true despite their theoretical difficulty (MIP is NP-hard, see e.g. [29]). Due to algorithmic developments and increased computational power, it is nowadays possible to solve problems of realistic size (at least for some problem classes).

For precisely the same reasons as for LP problems, it may be the case that some or all parameters of MIP problems are subject to uncertainty. Below we follow the approach of stochastic linear programming to handle this uncertainty, leading to mixed-integer recourse models. As we will see, this is a straightforward extension of the linear recourse models presented in Chap. 3.

Suppose that the following mixed-integer programming problem with random parameters in the constraints has been formulated:

$$(\mathrm{MIP}_0(\omega)) \qquad \min\left\{ cx : \begin{array}{c} Ax = b \\ T(\omega)x \sim h(\omega) \\ x \in X \end{array} \right\}$$

where the set X not only describes lower and upper bounds on the decision variables, but also specifies their type:

$$X = \{x \in \mathbb{Z}^{\bar{n}} \times \mathbb{R}^{n-\bar{n}} : x^l \leq x \leq x^u\},$$

i.e., the first $\bar{n}$ elements of the n-dimensional vector x are integer variables. As with the similar problem $\mathrm{LP}_0(\omega)$ in Sect. 3.2, default is $x^l = 0$ and $x^u = \infty$ in which case $X = \mathbb{Z}_+^{\bar{n}} \times \mathbb{R}_+^{n-\bar{n}}$. The constraints $Ax = b$ represent m_1 deterministic equality constraints, and $T(\omega)x \sim h(\omega)$ represent m random (in-)equality constraints. Here, $T(\omega)$ is an $m \times n$ matrix, and $h(\omega)$ is an $m \times 1$ vector, both depending linearly on a random vector $\omega \in \mathbb{R}^r$.

We consider $\mathrm{MIP}_0(\omega)$ as a decision problem, that is, we have to decide on x before knowing the actual value of ω which is only known in distribution. As we have seen in Chap. 3, one of the ways to obtain a meaningful model is to extend $\mathrm{MIP}_0(\omega)$ with a recourse structure (Y, q, W). As before, q is a $1 \times p$ vector of unit recourse costs and W is the $m \times p$ recourse matrix. As in the definition of the set X, simple bounds and variable type restrictions are represented by the set Y, defined as

$$Y = \{y \in \mathbb{Z}^{\bar{p}} \times \mathbb{R}^{p-\bar{p}} : y^l \leq y \leq y^u\},$$

so that the first $\bar{p}$ second-stage variables are integers. Usually, $Y = \mathbb{Z}_+^{\bar{p}} \times \mathbb{R}_+^{p-\bar{p}}$. Applying this recourse structure to $\text{MIP}_0(\omega)$ leads to the following mixed-integer recourse problem, which is a well-defined decision problem:

$$\text{(MIR)} \qquad \min_{x \in X} \left\{ cx + \mathbb{E}_\omega \Big[\underbrace{\min_{y \in Y} \{qy : Wy \sim h(\omega) - T(\omega)x\}}_{\text{second-stage MIP}} \Big] : Ax = b \right\}$$

The relation $\sim$ in the second stage is the same relation as in the random constraints of $\text{MIP}_0(\omega)$.

We see that the only differences between the mixed-integer recourse model MIR and the continuous recourse model SLPwR of the previous chapter are the variable type restrictions in X and Y. Hence, the alternative representations for SLPwR given in Sect. 3.2.1 carry over directly to MIR. For example, a compact representation is

$$\min_{x \in X} \{cx + Q(x) : Ax = b\},$$

where

$$Q(x) := \mathbb{E}_\omega [v(h(\omega) - T(\omega)x)], \quad x \in \mathbb{R}^n,$$

is the expected minimum recourse cost function, and

$$v(z) := \min_{y \in Y} \{qy : Wy \sim z\}, \quad z \in \mathbb{R}^m,$$

is the recourse penalty cost function. Note that v is now the *value function of a mixed-integer programming problem.*

Recall that if Ω is finite, it is possible to write SLPwR as an equivalent large scale linear programming problem. In the same way, MIR with discretely distributed parameters can be represented as a large scale mixed-integer programming problem. Obviously, the size of the resulting MIP makes it almost always impossible to solve it by standard methods. However, as we will see in Sect. 4.3 this large scale formulation is the starting point for several algorithms.

For later reference, when we discuss theoretical results and algorithms, we conclude the list of representations of the model MIR with the *canonical form*, in which all constraints are either equality constraints or nonnegativity constraints:

$$\min_{\tilde{x} \geq 0} \left\{ \tilde{c}\tilde{x} + \mathbb{E}_\omega \left[\min_{\tilde{y} \geq 0} \left\{ \tilde{q}\tilde{y} : \tilde{W}\tilde{y} = \tilde{h}(\omega) - \tilde{T}(\omega)\tilde{x} \right\} \right] : \tilde{A}\tilde{x} = \tilde{b} \right\} \tag{4.1}$$

Any MIR can be represented in such a canonical form (see Sect. 3.2.1).

The case study "Electricity Distribution" in Sect. 8.8 describes two examples of a decision problem modeled by a two-stage mixed-integer recourse model. In both problems, binary variables are needed to model the supply of small generators. Such generators can either supply at full capacity or not at all, so that the decision to use such a small generator corresponds to an on/off decision.

Remark 4.1.1 Throughout this chapter we assume that there exists an optimal solution to any MIP problem that has a finite optimal objective value. To guarantee this for the second-stage MIP, we assume that the recourse matrix W has *rational* elements only; see [28], also for a counterexample. From a practical perspective this assumption is not very restrictive.

4.1.1 Special Mixed-Integer Recourse Structures

Throughout this chapter we assume that the recourse structure (Y, q, W) is fixed, that is, we assume that Y, q and W do not depend on the random vector ω. This assumption is not necessary, but no theoretical results are known for models with non-deterministic recourse.

As was already discussed in the setting of continuous recourse models, the choice of a recourse structure for a problem $\mathrm{MIP}_0(\omega)$ is determined by the decision maker's interpretation of the problem as well as theoretical considerations. From the latter perspective, it is desirable to avoid infinite values of the recourse penalty function v. Since v is the value function of the second-stage mixed-integer problem given by the recourse structure (Y, q, W), this leads to conditions on the recourse structure that guarantee that v is finite everywhere. We repeat the definitions here for easy reference.

Definition 4.1.2 The fixed recourse structure (Y, q, W) is called *relatively complete* (i.e. relative with respect to $\mathrm{MIP}_0(\omega)$) if for all $\omega \in \Omega$ and all $x \in X$ with $Ax = b$ there exists a $y \in Y$, possibly depending on ω and x, such that $Wy \sim h(\omega) - T(\omega)x$. Equivalently, if $v(h(\omega) - T(\omega)x) < +\infty$ for all $\omega \in \Omega$ and $x \in \{x \in X : Ax = b\}$.

Definition 4.1.3 The fixed recourse structure (Y, q, W) is called *complete* if for all $z \in \mathbb{R}^m$ there exists a $y \in Y$, possibly depending on z, such that $Wy \sim z$. Equivalently, if $v(z) < +\infty$ for all $z \in \mathbb{R}^m$.

Although, formally, these definitions are the same as those for the continuous recourse case, there are differences, since here the sets X and/or Y contain integrality constraints.

Remark 4.1.4 Recall that for SLPwR in canonical form, completeness is determined by the recourse matrix $\tilde{W}$ only. Due to integrality restrictions, this is in general not true for mixed-integer recourse models. Indeed, even if $\tilde{W}$ is a complete recourse matrix in the 'continuous' sense, so that $\tilde{W}\tilde{y} = z$ has a solution in $\mathbb{R}^{\tilde{p}}_+$

for each $z \in \mathbb{R}^m$, there may not exist a mixed-integer solution in $\mathbb{Z}_+^{\bar{p}} \times \mathbb{R}_+^{\tilde{p}-\bar{p}}$. On the other hand, it is obvious that the recourse structure $(\tilde{Y}, \tilde{q}, \tilde{W})$ is complete if the submatrix corresponding to the continuous recourse variables is a crm.

If the recourse is not relatively complete, then the restricted recourse possibilities give rise to so-called induced constraints (see Sect. 3.2.3).

Definition 4.1.5 The fixed recourse structure (Y, q, W) is called *extremely inexpensive* if there exists a $z \in \mathbb{R}^m$ such that $v(z) = -\infty$.

Obviously, extremely inexpensive recourse structures are not suitable. Usually, they are avoided by assuming $q \geq 0$.

Next, we define the integer analogue of the *simple recourse* structure given in Definition 3.2.9. Similar as for continuous recourse models this structure penalizes all one-sided deviations from the goal constraints $z = 0$ separately. The penalty costs are calculated over the *rounded* surpluses and shortages, leading to the following *simple integer recourse* penalty function:

$$v(z) = \sum_{i=1}^{m} \left(q_i^+ \lceil z_i \rceil^+ + q_i \lfloor z_i \rfloor^- \right), \quad z \in \mathbb{R}^n.$$

To represent this second-stage value function v as a minimization problem in canonical form, we use the equivalent system of goal constraints $0 \geq z, 0 \geq -z$.

Definition 4.1.6 A fixed recourse structure, applicable to the system $0 \geq z$, $0 \geq -z$, $z \in \mathbb{R}^m$, is called *simple integer recourse* if

(i) $\tilde{Y} = \mathbb{Z}_+^{2m} \times \mathbb{R}_+^{2m}$, with $\tilde{Y} = \left\{(y^+, y^-, \tilde{y}^+, \tilde{y}^-) : y^+, y^- \in \mathbb{Z}_+^m,\ \tilde{y}^+, \tilde{y}^- \in \mathbb{R}_+^m\right\}$, and
(ii) its canonical form can be written as

$$\begin{pmatrix} \tilde{q} \\ \tilde{W} \end{pmatrix} = \begin{pmatrix} q^+ & q^- & 0 & 0 \\ I & 0 & -I & 0 \\ 0 & I & 0 & -I \end{pmatrix} \tag{4.2}$$

where q^+ and q^- are $1 \times m$ unit cost vectors and I denotes the $m \times m$ identity matrix.

In $\tilde{Y}$, the first two sets of variables represent the rounded deviations that are penalized, whereas the last two sets of variables represent the slack variables needed for the canonical form. For more details, we refer to Sect. 4.2.2, where simple integer recourse models will be discussed extensively.

4.2 Properties of Mixed-Integer Recourse Models

Compared to continuous recourse models, properties of mixed-integer recourse models are determined by two additional sources of possible difficulties: integrality restrictions in the first and in the second stage. If there are integer variables only in the first stage, then MIR is a mixed-integer problem with a convex objective function, since then the expected minimum recourse cost function Q is convex, similar as for continuous recourse problems in Chap. 3. Thus, aside from problems due to evaluating Q, its properties are the same as those of a deterministic mixed-integer problem. If on the other hand (some of) the *second-stage variables are integral*, then Q is the expectation of the *value function of a linear mixed-integer programming problem*, which has fundamentally different properties than its continuous counterpart.

In this section we discuss several properties of the function Q with integrality constraints in the second-stage variables. First we consider the fixed complete recourse model, and then in a separate section we study the pure integer simple recourse model.

4.2.1 *Fixed Mixed-Integer Recourse*

The results in this section are due to Schultz [39–41]. Their proofs are rather technical, and are therefore omitted. However, it is clear that properties of Q depend on properties of the value function v of the second-stage mixed-integer problem. Therefore, we first present some well-known properties of v.

Lemma 4.2.1 *Consider the function* v,

$$v(z) = \min_{y \in Y}\{qy : Wy = z\}, \quad z \in \mathbb{R}^m,$$

where $Y = \mathbb{Z}_+^{\bar{p}} \times \mathbb{R}_+^{p-\bar{p}}$, *the optimal value function of a MIP problem in canonical form.*

(i) The function v *is lower semicontinuous on* $\mathbb{R}^m$
(ii) The discontinuity points of v *are contained in a countable union of hyperplanes in* $\mathbb{R}^m$

If the recourse structure (Y, W, q) *is complete and sufficiently expensive then*

(iii) v *is finite on* $\mathbb{R}^m$. □

From an optimization point of view, most striking in this lemma is that convexity of the function v is lacking (at least if integrality plays a nontrivial role). Indeed, if the recourse structure is complete and sufficiently expensive, v is finite and discontinuous and therefore nonconvex.

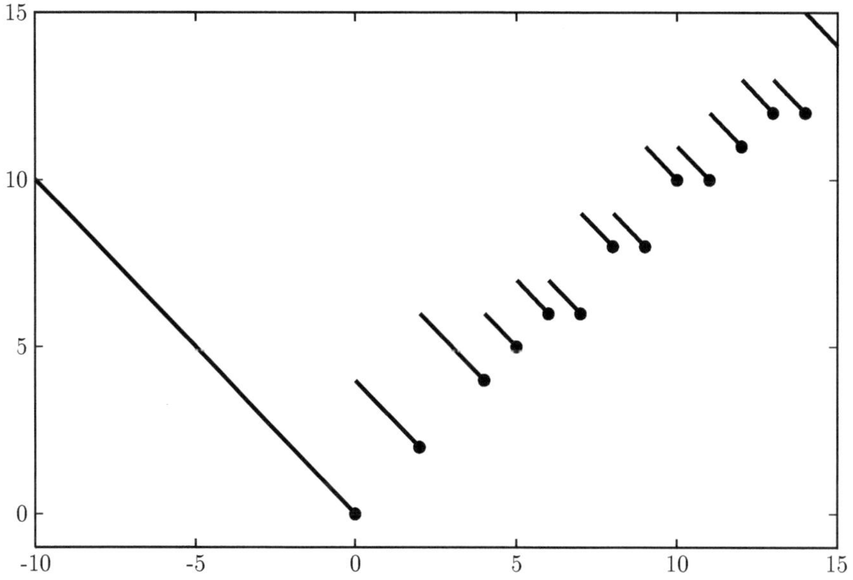

Fig. 4.1 The mixed-integer value function v of Example 4.2.2. The function value in discontinuity points is depicted as •

Example 4.2.2 Consider a one-dimensional example: for $z \in \mathbb{R}$,

$$v(z) = \min_{y \in Y} \left\{2y_1 + 5y_2 + 6y_3 + y_4 : 2y_1 + 5y_2 + 7y_3 - y_4 = z\right\},$$

where $Y = \mathbb{Z}^3_+ \times \mathbb{R}_+$. See Fig. 4.1 for the graph of v. In particular, note that the discontinuity points of v are contained in $\mathbb{Z}_+$, which is a countable union of hyperplanes in $\mathbb{R}$. ◇

Next we consider properties of the function $Q(x) = \mathbb{E}_\omega [v(h(\omega) - T(\omega)x)]$, $x \in \mathbb{R}^n$. It appears that the sets

$$D(x) := \left\{\omega \in \Omega : v \text{ is discontinuous in } h(\omega) - T(\omega)x\right\}, \quad x \in \mathbb{R}^n,$$

are important.

Theorem 4.2.3 *Consider a MIR model with fixed recourse structure (Y, q, W). If*

(a) the recourse structure is complete
(b) the recourse structure is not extremely inexpensive
(c) $\mathbb{E}_\omega [|\omega_i|] < \infty$, $i = 1, \ldots, r$

then $Q(x)$ is a finite lower semicontinuous function on $\mathbb{R}^n$. Moreover, Q is continuous at x if $\Pr\{\omega \in D(x)\} = 0$. □

We see that lower semicontinuity of Q is inherited from v, and that Q is finite on $\mathbb{R}^n$ if v is (conditions (a) and (b)), and in addition the same condition (c) holds that is familiar from the continuous recourse setting. Loosely speaking, the condition for continuity of Q at a fixed point x means that the set of ω such that v is discontinuous at $h(\omega) - T(\omega)x$ is negligible in the integral $\mathbb{E}_\omega[v(h(\omega) - T(\omega)x)]$; continuity of Q at x then follows since v is continuous at all $\omega \in \Omega \setminus D(x)$.

It follows from Theorem 4.2.3 that Q is continuous on $\mathbb{R}^n$ if $\Pr\{\omega \in D(x)\} = 0$ for all $x \in \mathbb{R}^n$.

Corollary 4.2.4 *If the conditional distribution of $h(\omega)$ given $T(\omega) = T$ is continuous for almost all T (with respect to the distribution of $T(\omega)$), then Q is continuous on $\mathbb{R}^n$.* □

For example, this condition is satisfied if $h(\omega)$ and $T(\omega)$ are independent (e.g., if T is fixed) and $h(\omega)$ is continuously distributed. Similarly, if the joint distribution of $h(\omega)$ and $T(\omega)$ is continuous, then Q is continuous on $\mathbb{R}^n$.

Example 4.2.5 Let the second-stage value function v be given as in Example 4.2.2 and assume that $T = 1$ is fixed and $h(\omega) = \omega$ is a one-dimensional random variable, so that $Q(x) = \mathbb{E}_\omega[v(\omega - x)]$, $x \in \mathbb{R}$. Since the discontinuity points of v are contained in $\mathbb{Z}_+$, it follows that $D(x) \subset \{\omega \in \Omega : \omega - x \in \mathbb{Z}_+\}$ is a countable set for all $x \in \mathbb{R}$. Hence, if ω is continuously distributed then $\Pr\{\omega \in D(x)\} = 0$ for all $x \in \mathbb{R}$ so that Q is continuous on $\mathbb{R}$. If ω is a discrete random variable then $\Pr\{\omega \in D(x)\} > 0$ if $x = \omega - k$ for some $\omega \in \Omega$ and $k \in \mathbb{Z}_+$, so that Q may be discontinuous in such a point x. See Fig. 4.2 for the graph of Q in case ω is uniformly distributed, respectively discrete on $\Omega = \{5, 5.25, 5.5, 5.75, 6\}$ (left) and continuous on $\Omega = [5, 6]$ (right).

◇

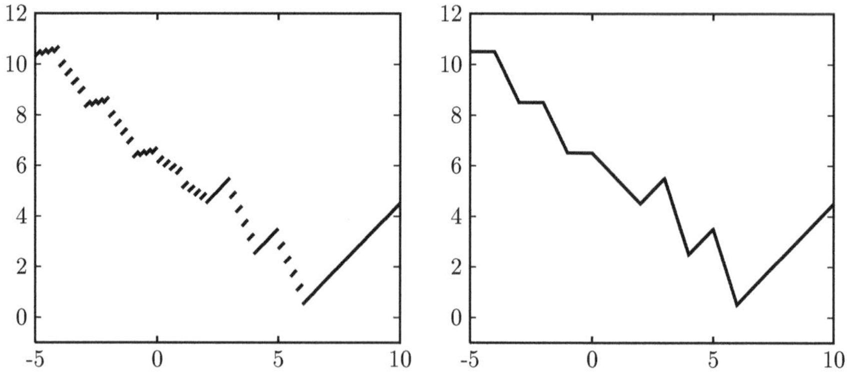

Fig. 4.2 The expected value function Q of Example 4.2.5 for $h(\omega)$ following a discrete (left) and continuous distribution (right)

It appears that in the proof of algorithms the concept of Lipschitz continuity is often useful. For the mixed-integer expected minimum recourse function Q, the conditions for Lipschitz continuity are given in the following theorem. They are rather technical. However, they are necessary as shown by Schultz in two examples [40]. We present the result for fixed T matrix; it can be extended to the random T case using conditional distributions.

Theorem 4.2.6 *Let $h(\omega)$ follow a continuous distribution. Assume that, for any non-singular linear transformation $B : \mathbb{R}^m \mapsto \mathbb{R}^m$, each one-dimensional marginal distribution of $Bh(\omega)$ has a bounded probability density function $f(s)$ that is monotone for $|s|$ sufficiently large. Then Q is Lipschitz continuous on any bounded subset of $\mathbb{R}^n$.*

For example, the (non-degenerate) multivariate normal distribution and the t-distribution satisfy the assumptions of this theorem.

To evaluate the expected value function at least approximately, it may be necessary to approximate the distribution of $h(\omega)$ and $T(\omega)$. This leads to the analysis of stability of optimal solutions and the optimal value under such approximations. In [39] Schultz shows that, under conditions which are not too restrictive in practice, the problem is stable in this sense if the approximating distributions converge weakly to the original distribution. In particular, this result justifies the use of discrete approximations of continuous distributions.

4.2.2 Simple Integer Recourse

Here we consider a special case of the fixed mixed-integer recourse model, known as (pure) simple integer recourse; see Definition 4.1.6. The canonical form of this recourse structure is characterized by $Y = \mathbb{Z}_+^{2m} \times \mathbb{R}_+^{2m}$ and

$$\begin{pmatrix} \tilde{q} \\ \tilde{W} \end{pmatrix} = \begin{pmatrix} q^+ & q^- & 0 & 0 \\ I & 0 & -I & 0 \\ 0 & I & 0 & -I \end{pmatrix}. \tag{4.3}$$

Recall that this recourse structure applies to $0 = z$, $z \in \mathbb{R}^m$, represented as the system $0 \geq z$, $0 \geq -z$, so that the second-stage problem has $2m$ rows.

This recourse structure is obviously complete, and it is not extremely inexpensive if $q^+ \geq 0$ and $q^- \geq 0$, as we will assume from now on.

Exercise 4.2.7 Prove this. ◁

The additional structure in this special case allows the derivation of further properties of the function Q, as we will show now.

It is clear from the recourse structure (4.3) that the continuous recourse variables $y^{2m+1}, \ldots, y^{4m}$ play the role of slack variables in this model. Thus, the second-stage problem can be written in inequality form as

$$v(z) = \min \left\{ q^+ y^+ + q^- y^- : \begin{array}{l} y^+ \geq z \\ y^- \geq -z \\ y^+ \in \mathbb{Z}_+^m,\ y^- \in \mathbb{Z}_+^m \end{array} \right\}.$$

Moreover, as a consequence of the simple recourse structure, the function v is separable: $v(z) = \sum_{i=1}^m v_i(z_i)$, where

$$v_i(z_i) = \min \left\{ q_i^+ y_i^+ + q_i^- y_i^- : \begin{array}{l} y_i^+ \geq z_i \\ y_i^- \geq -z_i \\ y_i^+ \in \mathbb{Z}_+,\ y_i^- \in \mathbb{Z}_+ \end{array} \right\}, \quad i = 1, \ldots, m.$$

Since $q_i^+ \geq 0$ and $q_i^- \geq 0$, as we assume, the minimization problem on the right has a trivial solution: if $z_i \geq 0$ then y_i^+ is the round up of z_i and $y_i^- = 0$; if $z_i < 0$ then $y_i^+ = 0$ and y_i^- is the round up of $-z_i$ (which is equal to minus the round down of z_i). We conclude that

$$v_i(z_i) = q_i^+ \lceil z_i \rceil^+ + q_i^- \lfloor z_i \rfloor^-, \quad z_i \in \mathbb{R}, \quad i = 1, \ldots, m,$$

where for $s \in \mathbb{R}$, $\lceil s \rceil^+ := \max\{0, \lceil s \rceil\}$ and $\lfloor s \rfloor^- := \max\{0, -\lfloor s \rfloor\} = \lceil -s \rceil^+$. Using this notation, the integer simple recourse model in canonical form can be written as

$$\min \left\{ cx + \sum_{i=1}^m \left(q_i^+ \mathbb{E}_\omega \left[\lceil h_i(\omega) - T_i(\omega)x \rceil^+ \right] + q_i^- \mathbb{E}_\omega \left[\lfloor h_i(\omega) - T_i(\omega)x \rfloor^- \right] \right) : \right.$$

$$\left. Ax = b,\ x \in \mathbb{Z}_+^{\bar{n}} \times \mathbb{R}_+^{n-\bar{n}} \right\}.$$

To guarantee finiteness of the expected recourse function Q we not only assume that $q^+ \geq 0$ and $q^- \geq 0$, but also that $\mathbb{E}_\omega [|\omega_i|] < \infty$, $i = 1, \ldots, r$, so that the conditions of Theorem 4.2.3 are satisfied.

We restrict the discussion to an important special case: *only the right-hand side is random*. That is,

$$h(\omega) := \omega \in \mathbb{R}^m$$

$$T(\omega) := T \quad \text{(deterministic)}$$

Introducing *tender variables* $s \in \mathbb{R}^m$, the simple integer recourse (SIR) model with fixed technology matrix T is

$$\min \left\{ cx + \sum_{i=1}^{m} Q_i(s_i) : \begin{array}{l} Ax \quad\;\; = b \\ Tx - s = 0 \\ x \in \mathbb{Z}_+^{\bar{n}} \times \mathbb{R}_+^{n-\bar{n}},\ s \in \mathbb{R}^m \end{array} \right\}$$

where

$$\begin{aligned} Q_i(s_i) &:= \mathbb{E}_{\omega_i}\left[v_i(\omega_i - s_i)\right] \\ &= q_i^+ \mathbb{E}_{\omega_i}\left[\lceil \omega_i - s_i \rceil^+\right] + q_i^- \mathbb{E}_{\omega_i}\left[\lfloor \omega_i - s_i \rfloor^-\right], \quad s_i \in \mathbb{R}. \end{aligned}$$

Below we will review properties of the so-called *one-dimensional simple integer recourse function* Q_i and its constituents,

$$\hat{G}_i(s_i) := \mathbb{E}_{\omega_i}\left[\lceil \omega_i - s_i \rceil^+\right], s_i \in \mathbb{R} : \ \text{the } \textit{integer expected surplus function},$$

$$\hat{H}_i(s_i) := \mathbb{E}_{\omega_i}\left[\lfloor \omega_i - s_i \rfloor^-\right], s_i \in \mathbb{R} : \ \text{the } \textit{integer expected shortage function},$$

so

$$Q_i(s_i) = q_i^+ \hat{G}_i(s_i) + q_i^- \hat{H}_i(s_i), \quad s_i \in \mathbb{R}, \quad i = 1, \ldots, m.$$

For a discussion of the names of the functions $\hat{G}_i$ and $\hat{H}_i$, see Remark 3.3.8.

Before going into details, we mention a special case that arises if in addition to the assumptions made above, it is also assumed that each ω_i has a finite discrete distribution. Then the simple integer recourse model with only right-hand side random can possibly be solved as a deterministic MIP problem, whose size is not too large in relation to the size of the underlying MIP model.

Exercise 4.2.8 Show this. (See also Exercise 3.3.5.) ◁

This observation is similar to that for analogous continuous simple recourse models. One should realize, however, that from a computational point of view the resulting MIP problem will be too large much sooner than the corresponding LP.

4.2.3 Integer Simple Recourse: Its One-Dimensional Functions

In this section we deal with basic functions for one-dimensional integer recourse, such as the one-dimensional integer expected shortage and surplus functions, the two components of the one-dimensional simple integer recourse function. These functions are obviously important in analyzing simple integer recourse models, but in addition they exhibit some of the characteristic properties of more general (mixed-)integer recourse functions. Thus, their study may also provide some insight and inspiration for the analysis of more general models.

The One-Dimensional Integer Value Function $\hat{v}$

First we consider properties of the underlying one-dimensional value function $\hat{v}$. For $z \in \mathbb{R}$,

$$\begin{aligned} v(z) &= \min \left\{ q^+y^+ + q^-y^- : \begin{array}{l} y^+ \geq z \\ y^- \geq -z \\ y^+ \in \mathbb{Z}_+,\ y^- \in \mathbb{Z}_+ \end{array} \right\} \\ &= q^+\lceil z \rceil^+ + q^-\lfloor z \rfloor^-, \end{aligned} \tag{4.4}$$

see Fig. 4.3. By Lemma 4.2.1, this value function $\hat{v}$ is finite and lower semicontinuous. Moreover, assuming that both $q^- > 0$ and $q^+ > 0$, it is easy to see that

(i) $\hat{v}$ is nonincreasing on $(-\infty, 0]$ and nondecreasing on $[0, \infty)$;
(ii) $\hat{v}$ is discontinuous at z if and only if $z \in \mathbb{Z}$. In particular, $\hat{v}$ is

- continuous from the right but not from the left at $z \in \mathbb{Z}_- \setminus \{0\}$, with $\lim_{s \uparrow z} \hat{v}(s) - \hat{v}(z) = q^-$;
- continuous from the left but not from the right at $z \in \mathbb{Z}_+ \setminus \{0\}$, with $\lim_{s \downarrow z} \hat{v}(s) - \hat{v}(z) = q^+$;
- discontinuous from both sides at $z = 0$, with $\hat{v}(0) = 0$ and $\lim_{s \uparrow 0} \hat{v}(s) - \hat{v}(0) = q^-$ and $\lim_{s \downarrow 0} \hat{v}(s) - \hat{v}(0) = q^+$.

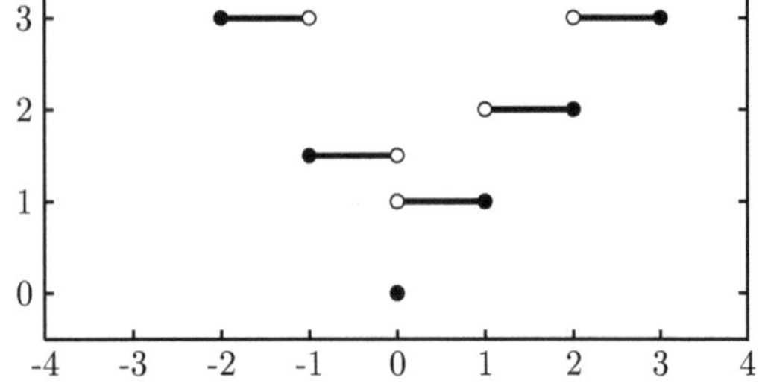

Fig. 4.3 The one-dimensional value function v with $q^+ = 1$, $q^- = 1.5$. The function value in discontinuity points is depicted as •

(iii) $\hat{v}$ is constant in between neighboring discontinuity points.
(iv) in comparison with its continuous counterpart $v(z) := q^+(z)^+ + q^-(z)^-$, $z \in \mathbb{R}$, it holds that $v(z) \leq \hat{v}(z) < v(z) + \max\{q^+, q^-\}$ for all $z \in \mathbb{R}$. In fact, $\hat{v}(z) = v(z)$ if $z \in \mathbb{Z}$.

The analogous properties of $\hat{v}$ for the case that either $q^- = 0$ or $q^+ = 0$ are obvious.

We see that $\hat{v}$ exhibits many properties of the general mixed-integer value function. At the same time, it is much more tractable due to the additional structure and, most importantly, the available closed form expression (4.4) is valid. Below we will see that this allows for more and sharper results compared to the general case.

As in the continuous simple recourse case, we introduce the integer versions of the expected surplus functions G and the expected shortage function H, defined in Chap. 3.

Definition 4.2.9 Let ω be a one-dimensional random variable. Then the one-dimensional *integer expected surplus function* is

$$\hat{G}(x) = \mathbb{E}_\omega\left[\lceil \omega - x \rceil^+\right], \qquad x \in \mathbb{R}.$$

Definition 4.2.10 Let ω be a one-dimensional random variable. Then the one-dimensional *integer expected shortage function* is

$$\hat{H}(x) = \mathbb{E}_\omega\left[\lfloor \omega - x \rfloor^-\right], \qquad x \in \mathbb{R}.$$

We start with explicit formulas for the functions $\hat{G}$ and $\hat{H}$, from which many properties of these functions can be derived.

$$\hat{G}(x) = \sum_{k=0}^{\infty} \Pr\{\omega > x + k\} = \sum_{k=0}^{\infty} \left(1 - F(x + k)\right) \tag{4.5}$$

$$\hat{H}(x) = \sum_{k=0}^{\infty} \Pr\{\omega < x - k\} = \sum_{k=0}^{\infty} \tilde{F}(x - k), \tag{4.6}$$

where $F(s) = \Pr\{\omega \leq s\}$ is the right-continuous cdf of ω and $\tilde{F}(s) = \Pr\{\omega < s\}$ is the left-continuous version of F, see Sect. A.3. We will only prove (4.5), the proof of (4.6) is similar and left as an exercise.

$$\begin{aligned}\hat{G}(x) &= \mathbb{E}_\omega\left[\lceil \omega - x \rceil^+\right] \\ &= \sum_{j=1}^{\infty} j \cdot \Pr\left\{\lceil \omega - x \rceil^+ = j\right\}.\end{aligned}$$

Using $j = \sum_{k=0}^{j-1} 1$ and then changing the order of summation we obtain

$$\begin{aligned}\hat{G}(x) &= \sum_{j=1}^{\infty}\sum_{k=0}^{j-1} \Pr\left\{\lceil \omega - x\rceil^{+} = j\right\} \\ &= \sum_{k=0}^{\infty}\sum_{j=k+1}^{\infty} \Pr\left\{\lceil \omega - x\rceil^{+} = j\right\} \\ &= \sum_{k=0}^{\infty} \Pr\left\{\lceil \omega - x\rceil^{+} > k\right\} \\ &= \sum_{k=0}^{\infty} \Pr\left\{\omega > x + k\right\}.\end{aligned}$$

There is an obvious similarity between the expressions (4.5)–(4.6) and those for their analogues G and H in continuous simple recourse, which we recall are

$$\begin{aligned}G(x) &= \int_x^{\infty} \big(1 - F(t)\big)\, dt = \int_x^{\infty} \Pr\{\omega \geq t\}\, dt \qquad (4.7)\\ H(x) &= \int_{-\infty}^{x} F(t)\, dt = \int_{-\infty}^{x} \Pr\{\omega \leq t\}\, dt.\end{aligned}$$

Exercise 4.2.11 Show that the formulae for the function $\hat{G}$ are direct consequences of (4.7). Hint: Apply (4.7) to $\mathbb{E}_{\eta}\left[\lceil \eta\rceil^{+}\right] = \mathbb{E}_{\eta}\left[(\lceil \eta\rceil - 0)^{+}\right]$, with η a random variable, and then substitute $\eta = \omega - x$ for any $x \in \mathbb{R}$. $\triangleleft$

We will now discuss the properties of the functions $\hat{G}$, $\hat{H}$, and $\hat{Q}$, in this order. In all these cases, we start with an informal description of the main issues, then a theorem is presented with precise descriptions. For complete proofs we refer to the literature. Unless stated otherwise, all results come from the Ph.D. thesis of van der Vlerk [47]. The properties of each of these functions are illustrated by means of numerical examples, with graphs.

The One-Dimensional Integer Expected Surplus Function $\hat{G}$

First consider the integer expected surplus function $\hat{G}$, defined by

$$\hat{G}(x) := \mathbb{E}_{\omega}\left[\lceil \omega - x\rceil^{+}\right], \quad x \in \mathbb{R}.$$

It differs from its continuous counterpart G, defined by $G(x) := \mathbb{E}_{\omega}\left[(\omega - x)^{+}\right]$, $x \in \mathbb{R}$, only by the round-up operation. Having this in mind, it is easy to see that $\hat{G}$ is a non-negative, non-increasing function, that is a little bit (less than 1) larger than G. The round-up operation causes $\hat{G}$ to be piecewise constant (hence non-continuous and non-convex) if the distribution of ω is discrete. Another property of the round-up function $r(z) := \lceil z \rceil$, $z \in \mathbb{R}$, is its 'periodicity': after each integer value, the function repeats itself more or less, up to a linear term. Such so-called semiperiodicity properties are inherited by $\hat{G}$, and these properties appear to be very useful for analyzing and approximating such functions. Due to the round-up operation, the restriction of $\hat{G}$ to $\mathbb{Z}$ (or any shift of the integer grid $\alpha + \mathbb{Z}$, for some $\alpha \in \mathbb{R}$) is a convex function. But unlike its continuous counterpart $\hat{G}$ is non-convex on the whole of $\mathbb{R}$, in general. Nevertheless, in some cases $\hat{G}$ is a convex function on $\mathbb{R}$. This can only be true if ω has a continuous distribution. It appears to be possible to characterize those pdf's that generate convex functions $\hat{G}$. This characterization is helpful for finding better convex approximations of $\hat{G}$ than G, and convex approximations are useful for the development of algorithms, of course. Concluding this discussion, we note that one-sided derivatives of $\hat{G}$ exist, if the pdf has bounded total variation.

In the next theorem a precise formulation of all these properties of $\hat{G}$ are given, together with the corresponding formulas.

Theorem 4.2.12 *Let ω be a random variable with cdf F and assume that $\mu^+ := \mathbb{E}_\omega\left[(\omega)^+\right] < +\infty$. Define the* integer expected surplus function $\hat{G}$ *as*

$$\hat{G}(x) = \mathbb{E}_\omega\left[\lceil \omega - x \rceil^+\right], \qquad x \in \mathbb{R}.$$

Then,

(a) $\hat{G}$ is finite and non-negative.
(b) $\hat{G}$ is non-increasing.
(c) For all $x \in \mathbb{R}$,

$$\hat{G}(x) = \sum_{k=0}^{\infty} \Pr\{\omega > x + k\} = \sum_{k=0}^{\infty} \bigl(1 - F(x+k)\bigr). \tag{4.8}$$

Moreover, the corresponding continuous expected surplus function G provides a convex lower bound as well as a convex upper bound for $\hat{G}$:

$$G(x) \le \hat{G}(x) \le G(x) + 1, \quad x \in \mathbb{R}.$$

A sharper upper bound is given by $\hat{G}(x) \le G(x) + 1 - F(x)$, $x \in \mathbb{R}$.
(d) **Semiperiodicity:** *Assume that* $\Pr\{\omega \in (a,b)\} = 0$, *where $-\infty \le a < b \le +\infty$. Then $\hat{G}$ is semi-periodic (see Definition B.41) with period* 1 *and slope $\gamma = F(a) - 1$ on the interval $[a, b+1) \cap \mathbb{R}$. That is, if $x_1, x_2 \in [a, b+1)$, $x_1 - x_2 = n \in \mathbb{Z}$ then*

$$\hat{G}(x_2) - \hat{G}(x_1) = n\gamma.$$

In particular, if $\Pr\{\omega < \underline{\omega}\} = 0$ *for some* $\underline{\omega} > -\infty$ *then* $\hat{G}$ *is semi-periodic with period* 1 *and slope* -1 *on* $(-\infty, \underline{\omega}]$.

(e) **Continuity and differentiability:** *Suppose that* ω *is* continuously distributed *with pdf* f. *Then* $\hat{G}$ *is continuous on* $\mathbb{R}$. *If the pdf* f *has finite total variation* $|\Delta| f$ *(see Definition A.3.1) then* $\hat{G}$ *is even* Lipschitz continuous *with a Lipschitz constant that is at most* $1 + |\Delta| f/2$. *In that case* $\hat{G}$ *has left and right derivatives everywhere, given by*

$$\hat{G}'_-(x) = -\sum_{k=0}^{\infty} f_-(x+k), \quad x \in \mathbb{R},$$

and

$$\hat{G}'_+(x) = -\sum_{k=0}^{\infty} f_+(x+k), \quad x \in \mathbb{R},$$

respectively, where $f_-(s) := \lim_{\varepsilon \uparrow 0} f(s+\varepsilon)$ *and* $f_+(s) := \lim_{\varepsilon \downarrow 0} f(s+\varepsilon)$ *are the left and right continuous versions of* f. *In particular,* $\hat{G}$ *is* differentiable *at* $x \in \mathbb{R}$ *if* f *is continuous at* $x+k$ *for all* $k \in \mathbb{Z}_+$, *and in that case the derivative is given by*

$$\hat{G}'(x) = -\sum_{k=0}^{\infty} f(x+k).$$

(f) **Convexity:** *For any shift parameter* $\alpha \in \mathbb{R}$, *the restriction of* $\hat{G}$ *to* $\{\alpha + \mathbb{Z}\}$ *is* convex. *Moreover, if we assume that* ω *has a regular pdf* $f \in \mathcal{R}$ *(see Sect. A.3), then the function* $\hat{G}$ *is convex on the whole of* $\mathbb{R}$ *if and only if* $f(s) = \Phi(s+1) - \Phi(s)$, $s \in \mathbb{R}$, *where* Φ *is an arbitrary cdf with finite mean value.*

(g) **Discrete distributions:** *Suppose that* ω *is discretely distributed with support* Ω. *Then the function* $\hat{G}$ *is lower semicontinuous and continuous from the right. The set of all discontinuity points of* $\hat{G}$ *is given by* $\Omega - \mathbb{Z}_+$. *If* $\hat{G}$ *is discontinuous at* x, *then the size of the discontinuity equals* $-\Pr\{\omega \in x + \mathbb{Z}_+\}$.

In between two successive points in $\Omega - \mathbb{Z}_+$ *the function* $\hat{G}$ *is constant. In particular, if* $\omega - a$ *is integer valued for some fixed scalar* a, *then the function* $\hat{G}$ *is constant on all intervals* $[a+n, a+n+1)$, $n \in \mathbb{Z}$. *Moreover, in that case* $\hat{G}(a+n) = G(a+n)$ *for all such* n.

Exercise 4.2.13 Prove that $\hat{G}$ is constant in between successive discontinuity points, if ω is discretely distributed. ◁

We conclude this discussion on $\hat{G}$ by means of two numerical examples, with graphs: the first one has a continuous random variable, whereas the second has a discrete one.

Example 4.2.14 Assume that ω follows an exponential distribution with parameter λ. Then, using the analytic expression for $\hat{G}$ in Theorem 4.2.12 (c), we have

$$\hat{G}(x) = \sum_{k=0}^{\infty} \Big(1 - F(x+k)\Big), \quad x \in \mathbb{R},$$

where $F(x+k) = 1 - e^{-\lambda(x+k)}$ if $x + k > 0$ and $F(x+k) = 0$, otherwise. Observe that if $x > 0$, then $x + k > 0$ for all $k \in \mathbb{Z}_+$, and thus

$$\hat{G}(x) = \sum_{k=0}^{\infty} e^{-\lambda(x+k)} = e^{-\lambda x} \sum_{k=0}^{\infty} (e^{-\lambda})^k = \frac{e^{-\lambda x}}{1 - e^{-\lambda}}, \quad x > 0.$$

On the other hand, if $x < 0$ with $x \notin \mathbb{Z}$, then $x + k < 0$ for $k = 0, \ldots, \lfloor -x \rfloor$, and thus

$$\hat{G}(x) = \sum_{k=0}^{\lfloor -x \rfloor} \Big(1 - F(x+k)\Big) + \sum_{k=\lceil -x \rceil}^{\infty} e^{-\lambda(x+k)} = \lceil -x \rceil + \frac{e^{-\lambda(x + \lceil -x \rceil)}}{1 - e^{-\lambda}}, \quad x < 0.$$

It is not hard to verify that the above expression also holds if $x \in \mathbb{Z}_-$. Hence, combining all results, we conclude that the following expression holds for all $x \in \mathbb{R}$:

$$\hat{G}(x) = \lceil -x \rceil^+ + \frac{e^{-\lambda(x + \lceil -x \rceil^+)}}{1 - e^{-\lambda}}, \quad x \in \mathbb{R}.$$

Figure 4.4 shows $\hat{G}$ for $\lambda = 3$. Clearly, $\hat{G}$ is non-negative and non-increasing. Moreover, since ω is continuously distributed, the function $\hat{G}$ is continuous.

Fig. 4.4 The one-dimensional integer expected surplus function $\hat{G}$ for $\omega \sim \mathcal{E}(3)$

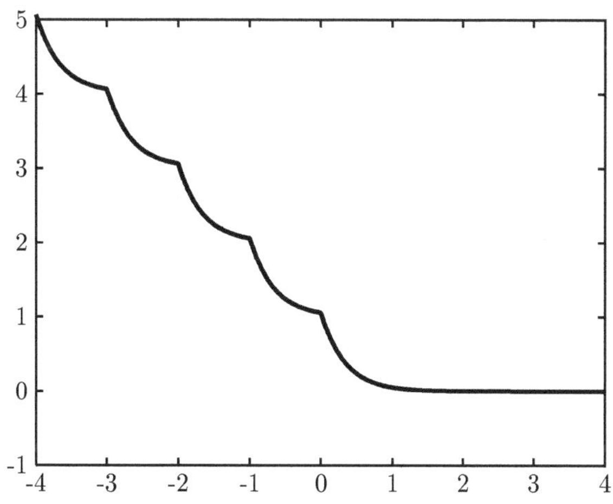

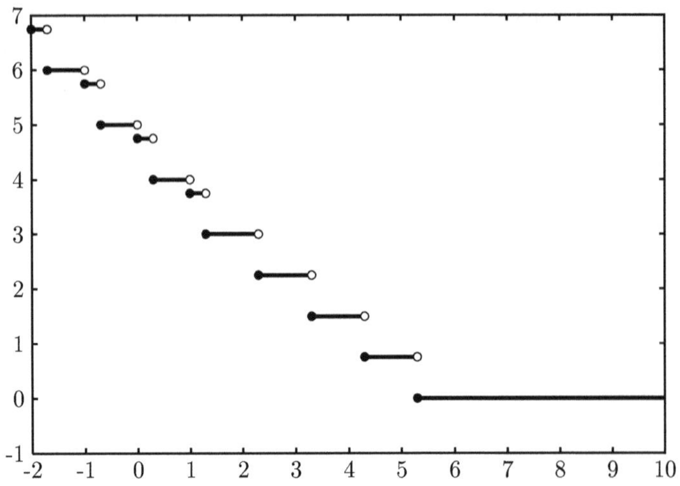

Fig. 4.5 The one-dimensional integer expected surplus function $\hat{G}$ for ω discretely distributed with $\Pr\{\omega = 1\} = 0.25$ and $\Pr\{\omega = 5.3\} = 0.75$

Furthermore, $\hat{G}$ is differentiable everywhere except for $x \in \mathbb{Z}_-$. Indeed, the exponential density function

$$f(x) = \begin{cases} \lambda e^{-\lambda x}, & x \geq 0, \\ 0, & \text{otherwise}, \end{cases}$$

is discontinuous at $x = 0$, which means that f is not continuous at $x + k$ for all $k \in \mathbb{Z}_+$ if and only if $x \in \mathbb{Z}_-$, see Theorem 4.2.12 (e). Finally, in line with Theorem 4.2.12 (d), $\hat{G}$ is semi-periodic with period 1 and slope $\gamma = -1$ on the interval $(-\infty, 0]$, since $\Pr\{\omega < 0\} = 0$. ◇

Example 4.2.15 Assume that ω is discretely distributed with $\Pr\{\omega = 1\} = 0.25$ and $\Pr\{\omega = 5.3\} = 0.75$, and consider the integer expected surplus function $\hat{G}$, see Fig. 4.5. Since ω is not continuously distributed, the function $\hat{G}$ is not continuous, but in line with Theorem 4.2.12 (g), the function $\hat{G}$ is lower semicontinuous and continuous from the right. Observe that $\hat{G}$ has discontinuity points at $x \in 1 - \mathbb{Z}_+$ and $x \in 5.3 - \mathbb{Z}_+$, and that $\hat{G}$ is constant in between two successive discontinuity points. Finally, in line with Theorem 4.2.12 (d), observe that $\hat{G}$ is semi-periodic on the intervals $(-\infty, 1)$, $[1, 5.3)$, and $[5.3, \infty)$, with period 1 and slopes $\gamma = -1, -0.75$, and 0, respectively. ◇

The One-Dimensional Integer Expected Shortage Function $\hat{H}$

We now start with the discussion of the properties of the integer expected *shortage* function $\hat{H}$, defined by $\hat{H}(x) := \mathbb{E}_\omega\left[\lfloor \omega - x \rfloor^-\right], x \in R$. In fact, since

$$\lfloor \omega - x \rfloor^- = \lfloor -(x-\omega) \rfloor^- = \lceil x - \omega \rceil^+ = \lceil (-\omega) - (-x) \rceil^+,$$

the function $\hat{H}$ is an elementary transformation (via $x \to -x$) of an elementary adjustment of $\hat{G}$ (via $\omega \to -\omega$). Hence, every property of $\hat{G}$ can be transformed into the related property of $\hat{H}$, by using the following lemma

Lemma 4.2.16 *Let ω be a random variable. Define $\zeta = -\omega$. Then*

$$\hat{H}(x) = \hat{G}^\zeta(-x), \quad x \in \mathbb{R},$$

where $\hat{G}^\zeta(x) = \mathbb{E}_\zeta\left[\lceil \zeta - x \rceil^+\right]$.

The random variable ζ has cdf $F_\zeta(s) = 1 - \tilde{F}(-s)$, where $\tilde{F}(s) = \Pr\{\omega < s\}$ is the left continuous cdf of ω. If ω has a pdf f then ζ has a pdf $f_\zeta(s) = f(-s)$. □

Exercise 4.2.17 Prove Lemma 4.2.16. Hint: Use that $\lfloor s \rfloor^- = \lceil -s \rceil^+,\ s \in \mathbb{R}$. ◁

So $\hat{H}$ is a non-negative, non-decreasing function, a little bit (less than 1) larger than its continuous counterpart H, defined as $H(x) := \mathbb{E}_\omega\left[(\omega - x)^-\right], x \in \mathbb{R}$. Just as $\hat{G}$, $\hat{H}$ is piecewise constant for discretely distributed ω, and continuous for continuous distributions of ω. It has similar semi-periodicity, continuity, differentiability, and convexity properties. Its formulas follow from those for $\hat{G}$, by applying the transformation $(x, \omega) \to (-x, -\omega)$.

For sake of completeness we give the complete analogue of Theorem 4.2.12 for $\hat{H}$.

Theorem 4.2.18 *Let ω be a random variable with left-continuous cdf $\tilde{F}$ and assume that $\mu^- := \mathbb{E}_\omega\left[(\omega)^-\right] < +\infty$. Define the* integer expected shortage function $\hat{H}$ *as*

$$\hat{H}(x) = \mathbb{E}_\omega\left[\lfloor \omega - x \rfloor^-\right], \qquad x \in \mathbb{R}.$$

Then,

(a) $\hat{H}$ is finite and non-negative.
(b) $\hat{H}$ is non-decreasing.
(c) For all $x \in \mathbb{R}$,

$$\hat{H}(x) = \sum_{k=0}^{\infty} \Pr\{\omega < x - k\} = \sum_{k=0}^{\infty} \tilde{F}(x-k), \tag{4.9}$$

where $\tilde{F}(s) = \Pr\{\omega < s\}$, $s \in \mathbb{R}$, is the left continuous cdf of ω.

Moreover, the corresponding continuous expected shortage function H provides a convex lower bound as well as a convex upper bound for $\hat{H}$:

$$H(x) \le \hat{H}(x) \le H(x) + 1, \quad x \in \mathbb{R}.$$

A sharper upper bound is given by $\hat{H}(x) \le H(x) + \tilde{F}(x)$, $x \in \mathbb{R}$.

(d) **Semiperiodicity:** *Assume that* $\Pr\{\omega \in (a, b)\} = 0$*, where* $-\infty \le a < b \le +\infty$*. Then $\hat{H}$ is semi-periodic (see Definition B.41) with period* 1 *and slope $\gamma = \hat{F}(b)$ on the interval $(a-1, b] \cap \mathbb{R}$. That is, if $x_1, x_2 \in (a-1, b]$, $x_1 - x_2 = n \in \mathbb{Z}$ then*

$$\hat{H}(x_1) - \hat{H}(x_2) = n\gamma.$$

In particular, if $\Pr\{\omega > \overline{\omega}\} = 0$ *for some $\overline{\omega} < \infty$ then $\hat{H}$ is semi-periodic with period* 1 *and slope* 1 *on $[\overline{\omega}, \infty)$.*

(e) **Continuity and differentiability:** *Suppose that ω is* continuously distributed with pdf f. *Then $\hat{H}$ is continuous on $\mathbb{R}$. If the pdf f has finite total variation $|\Delta| f$ (see Definition A.3.1) then $\hat{H}$ is even Lipschitz continuous with a Lipschitz constant that is at most* $1 + |\Delta| f/2$. *In that case $\hat{H}$ has left and right derivatives everywhere, given by*

$$\hat{H}'_-(x) = \sum_{k=0}^{\infty} f_-(x-k), \quad x \in \mathbb{R},$$

and

$$\hat{H}'_+(x) = \sum_{k=0}^{\infty} f_+(x-k), \quad x \in \mathbb{R},$$

respectively, where $f_-(s) := \lim_{\varepsilon \uparrow 0} f(s+\varepsilon)$ *and* $f_+(s) := \lim_{\varepsilon \downarrow 0} f(s+\varepsilon)$*, that is, f_+ and f_- are the left and right continuous versions of f. In particular, $\hat{H}$ is differentiable at $x \in \mathbb{R}$ if f is continuous at $x - k$ for all $k \in \mathbb{Z}_+$, and in that case the derivative is given by*

$$\hat{H}'(x) = \sum_{k=0}^{\infty} f(x-k).$$

(f) **Convexity:** *For any shift parameter $\alpha \in \mathbb{R}$, the restriction of $\hat{H}$ to $\{\alpha + \mathbb{Z}\}$ is convex. Moreover, if we assume that ω has a regular pdf $f \in \mathcal{R}$ (see Sect. A.3), then the function $\hat{H}$ is convex on the whole of $\mathbb{R}$ if and only if $f(s) = \Phi(s+1) - \Phi(s)$, $s \in \mathbb{R}$, where Φ is an arbitrary cdf with finite mean value.*

(g) **Discrete distributions:** *Suppose that ω is discretely distributed with support Ω. Then the function $\hat{H}$ is lower semicontinuous and continuous from the left. The*

set of all discontinuity points of $\hat{H}$ *is given by* $\Omega + \mathbb{Z}_+$. *If* $\hat{H}$ *is discontinuous at* x, *then the size of the discontinuity equals* $\Pr\{\omega \in x - \mathbb{Z}_+\}$.

In between two successive points in $\Omega + \mathbb{Z}_+$ *the function* $\hat{H}$ *is constant. In particular, if* $\omega - a$ *is integer valued for some fixed scalar* a, *then the function* $\hat{H}$ *is constant on all intervals* $(a+n, a+n+1]$, $n \in \mathbb{Z}$. *Moreover, in that case* $\hat{H}(a+n) = H(a+n)$ *for all such* n. □

We conclude the discussion on $\hat{H}$ by means of two numerical examples, with graphs. The underlying distributions of ω are the same as in the examples for $\hat{G}$.

Example 4.2.19 Assume that ω follows an exponential distribution with parameter λ. Then, using the analytic expression for $\hat{H}$ in Theorem 4.2.18 (c), we have

$$\hat{H}(x) = \sum_{k=0}^{\infty} \tilde{F}(x-k), \quad x \in \mathbb{R},$$

where $\tilde{F}(x-k) = 1 - e^{-\lambda(x-k)}$ if $x - k > 0$ and $F(x-k) = 0$, otherwise. Observe that if $x < 0$, then $x - k < 0$ for all $k \in \mathbb{Z}_+$, and thus $\hat{H}(x) = 0$. On the other hand, if $x > 0$ with $x \notin \mathbb{Z}$, then $x - k > 0$ for $k = 0, \ldots, \lfloor x \rfloor$, and thus

$$\hat{H}(x) = \sum_{k=0}^{\lfloor x \rfloor} \tilde{F}(x-k) = \sum_{k=0}^{\lfloor x \rfloor} \left(1 - e^{-\lambda(x-k)}\right) = \lceil x \rceil - e^{-\lambda x} \frac{1 - e^{\lambda \lceil x \rceil}}{1 - e^{\lambda}}, \quad x > 0.$$

Since this expression also holds for $x \in \mathbb{Z}_+$, combining the above results, we conclude that

$$\hat{H}(x) = \lceil x \rceil^+ - e^{-\lambda x} \frac{1 - e^{\lambda \lceil x \rceil^+}}{1 - e^{\lambda}}, \quad x \in \mathbb{R}.$$

Figure 4.6 shows $\hat{H}$ for $\lambda = 3$. Clearly, $\hat{H}$ is non-negative and non-decreasing. Moreover, since ω is continuously distributed, the function $\hat{H}$ is continuous. Furthermore, $\hat{H}$ is differentiable everywhere except for $x \in \mathbb{Z}_+$. Indeed, the exponential density function

$$f(x) = \begin{cases} \lambda e^{-\lambda x}, & x \geq 0, \\ 0, & \text{otherwise}, \end{cases}$$

is discontinuous at $x = 0$, which means that f is not continuous at $x - k$ for all $k \in \mathbb{Z}_+$ if and only if $x \in \mathbb{Z}_+$, see Theorem 4.2.18 (e). Finally, in line with Theorem 4.2.18 (d), $\hat{H}$ is semi-periodic with period 1 and slope $\gamma = 0$ on the interval $(-\infty, 0]$, since $\Pr\{\omega < 0\} = 0$. ◇

Example 4.2.20 Assume that ω is discretely distributed with $\Pr\{\omega = 1\} = 0.25$ and $\Pr\{\omega = 5.3\} = 0.75$, and consider the integer expected shortage function $\hat{H}$, see

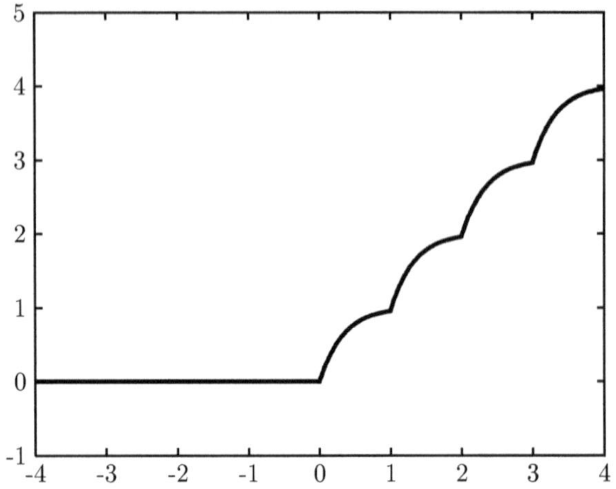

Fig. 4.6 The one-dimensional integer expected shortage function $\hat{H}$ for $\omega \sim \mathcal{E}(3)$

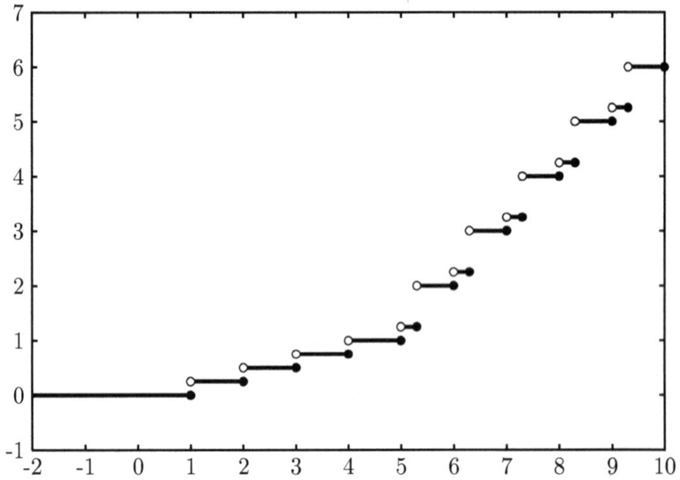

Fig. 4.7 The one-dimensional integer expected shortage function $\hat{H}$ for ω discretely distributed with $\Pr\{\omega = 1\} = 0.25$ and $\Pr\{\omega = 5.3\} = 0.75$

Fig. 4.7. Since ω is not continuously distributed, the function $\hat{H}$ is not continuous, but in line with Theorem 4.2.18 (g), the function $\hat{H}$ is lower semicontinuous and continuous from the left. Observe that $\hat{H}$ has discontinuity points at $x \in 1 + \mathbb{Z}_+$ and $x \in 5.3 + \mathbb{Z}_+$, and that $\hat{H}$ is constant in between two successive discontinuity points. Finally, in line with Theorem 4.2.18 (d), observe that $\hat{H}$ is semi-periodic on the intervals $(-\infty, 1)$, $[1, 5.3)$, and $[5.3, \infty)$, with period 1 and slopes $\gamma = 0, 0.25$, and 1, respectively.

$\diamond$

The One-Dimensional Integer Expected Value Function $\hat{Q}$

Finally, we are ready to state the properties of the generic *one-dimensional integer expected optimal value function* $\hat{Q}$, defined by

$$\hat{Q}(x) = \mathbb{E}_\omega \left[\min \left\{ q^+y^+ + q^-y^- : \begin{array}{l} y^+ \geq \omega - x \\ y^- \geq x - \omega \\ y^+ \in \mathbb{Z}_+,\ y^- \in \mathbb{Z}_+ \end{array} \right\} \right], \quad x \in \mathbb{R},$$

where ω is a random variable and $q^+ \geq 0$ and $q^- \geq 0$ are two scalars. Since

$$\hat{Q}(x) = q^+ \cdot \hat{G}(x) + q^- \cdot \hat{H}(x), \quad x \in \mathbb{R},$$

the properties of $\hat{Q}$ follow directly from those of $\hat{G}$ and $\hat{H}$.

In the theorem below we assume that q^+ and q^- are both strictly positive, since otherwise $\hat{Q}$ equals a multiple of one of the functions $\hat{G}$ or $\hat{H}$ (or is identically zero). It is also assumed that $\mu := \mathbb{E}_\omega[\omega]$ is finite, so that $\hat{Q}(x)$ is finite for all $x \in \mathbb{R}$. Similar as for the functions $\hat{G}$ and $\hat{H}$, the function $\hat{Q}$ is finite and non-negative. Its value is a little bit (at most $\max\{q^+, q^-\}$) larger than that of its continuous counterpart Q. As the sum of a non-increasing and non-decreasing function, its value can go up and down, depending on which term is dominant. It has semiperiodicity properties. In case ω is discretely distributed, $\hat{Q}$ is piecewise constant, and its discontinuities can be described explicitly. In case ω is continuously distributed, $\hat{Q}$ is a continuous function. If furthermore its pdf f has finite total variation, then $\hat{Q}$ is even Lipschitz continuous. In that case, $\hat{Q}$ has left and right derivatives everywhere, and they can be expressed in terms of f. Just as $\hat{G}$ and $\hat{H}$, $\hat{Q}$ is convex on any grid of the type $\alpha + \mathbb{Z}$, for a fixed $\alpha \in \mathbb{R}$. Precise conditions on the pdf f are formulated to guarantee that $\hat{Q}$ is convex on the whole of $\mathbb{R}$, actually.

All these properties are formulated precisely in the next theorem, an analogue of Theorems 4.2.12 and 4.2.18 for $\hat{G}$ and $\hat{H}$, respectively. It is also assumed that $\mathbb{E}_\omega[|\omega|] < +\infty$, implying that $\mu := \mathbb{E}_\omega[\omega]$ is finite, $\mathbb{E}_\omega\left[(\omega)^+\right] < +\infty$ and $\mathbb{E}_\omega\left[(\omega)^-\right] < +\infty$.

Theorem 4.2.21 *Let $\hat{Q}$ be the* one-dimensional simple integer recourse function, *defined as*

$$\hat{Q}(x) = q^+\mathbb{E}_\omega\left[\lceil \omega - x \rceil^+\right] + q^-\mathbb{E}_\omega\left[\lfloor \omega - x \rfloor^-\right], \quad x \in \mathbb{R},$$

where q^+ and q^- are positive scalars, and ω is a random variable with right continuous cdf F and left continuous cdf $\tilde{F}$, and with $\mathbb{E}_\omega[|\omega|] < +\infty$. Then

(a) For all $x \in \mathbb{R}$, $\hat{Q}(x)$ is finite and non-negative, and

$$\hat{Q}(x) = q^+ \sum_{k=0}^{\infty} \Pr\{\omega > x + k\} + q^- \sum_{k=0}^{\infty} \Pr\{\omega < x - k\}. \qquad (4.10)$$

Moreover, the corresponding one-dimensional continuous simple recourse expected value function Q, given by

$$Q(x) = q^+\mathbb{E}_\omega\left[(\omega - x)^+\right] + q^-\mathbb{E}_\omega\left[(\omega - x)^-\right], \quad x \in \mathbb{R},$$

provides a convex lower bound as well as a convex upper bound for $\hat{Q}$:

$$Q(x) \leq \hat{Q}(x) \leq Q(x) + \max\{q^+, q^-\}, \quad x \in \mathbb{R}.$$

(b) **Semiperiodicity:** *If $(a, b) \subset \mathbb{R}$ is an interval of length more than one such that $\Pr\{a < \omega < b\} = 0$, then $\hat{Q}$ is semi-periodic with period 1 and slope $\gamma = -q^+ \Pr\{\omega > a\} + q^- \Pr\{\omega < b\}$ on the interval $[a, b] \cap \mathbb{R}$. That is, if $x_1 \in [a, b]$, $x_2 \in [a, b]$, $x_1 - x_2 = n \in \mathbb{Z}$ then*

$$\hat{Q}(x_2) - \hat{Q}(x_1) = n\gamma.$$

In particular, if $\Pr\left\{\omega < \underline{\omega}\right\} = 0$ for some $\underline{\omega} > -\infty$ then $\hat{Q}$ is semi-periodic with period 1 and slope $-q^+$ on $(-\infty, \underline{\omega}]$. Similarly, if $\Pr\{\omega > \overline{\omega}\} = 0$ for some $\overline{\omega} < +\infty$ then $\hat{Q}$ is semi-periodic with period 1 and slope q^- on $[\overline{\omega}, \infty)$.

(c) **Continuity and differentiability:** *Suppose that ω is* continuously distributed *with pdf f. Then $\hat{Q}$ is continuous on $\mathbb{R}$. If the pdf f has a finite total variation $|\Delta| f$ then $\hat{Q}$ is even Lipschitz continuous with a Lipschitz constant that is at most* $\max\{q^+, q^-\}(1 + |\Delta| f/2)$. *In that case $\hat{Q}$ has left and right derivatives everywhere, given by*

$$\hat{Q}'_-(x) = -q^+ \sum_{k=0}^{\infty} f_-(x + k) + q^- \sum_{k=0}^{\infty} f_-(x - k), \quad x \in \mathbb{R},$$

and

$$\hat{Q}'_+(x) = -q^+ \sum_{k=0}^{\infty} f_+(x + k) + q^- \sum_{k=0}^{\infty} f_+(x - k), \quad x \in \mathbb{R},$$

respectively, where f_- and f_+ denote the left and right continuous version of f, respectively. In particular, $\hat{Q}$ is differentiable at $x \in \mathbb{R}$ if f is continuous at $x + k$ for all $k \in \mathbb{Z}$, and in that case the derivative is given by

$$\hat{Q}'(x) = -q^+ \sum_{k=0}^{\infty} f(x + k) + q^- \sum_{k=0}^{\infty} f(x - k), \quad x \in \mathbb{R}.$$

(d) **Convexity:** *For any shift parameter $\alpha \in \mathbb{R}$, the restriction of $\hat{Q}$ to $\{\alpha + \mathbb{Z}\}$ is convex. Moreover, if we assume that ω has a regular pdf $f \in \mathcal{R}$ (see Sect. A.3),* *then the function $\hat{Q}$ is convex on the whole of $\mathbb{R}$ if and only if $f(s) = \Phi(s+1) - \Phi(s)$, $s \in \mathbb{R}$, where Φ is an arbitrary cdf with finite mean value.*

(e) **Discrete distributions:** *Suppose that ω is* discretely distributed *with support Ω. Then the function $\hat{Q}$ is lower semicontinuous, and the set of all discontinuity points of $\hat{Q}$ is given by $\Omega + \mathbb{Z}$. In particular,*

(i) $\hat{Q}$ is continuous from the left but not from the right in all points of $(\Omega+\mathbb{Z}) \setminus (\Omega - \mathbb{Z}_+)$. If $\bar{x}$ is such a discontinuity point then $\lim_{x\downarrow\bar{x}} \hat{Q}(x) - \hat{Q}(\bar{x}) = q^- \Pr\{\omega \in \bar{x} - \mathbb{Z}_+\}$.

(ii) $\hat{Q}$ is continuous from the right but not from the left in all points of $(\Omega-\mathbb{Z}) \setminus (\Omega + \mathbb{Z}_+)$. If $\bar{x}$ is such a discontinuity point then $\hat{Q}(\bar{x}) - \lim_{x\uparrow\bar{x}} \hat{Q}(x) = -q^+ \Pr\{\omega \in \bar{x} + \mathbb{Z}_+\}$.

(iii) $\hat{Q}$ is neither continuous from the left nor continuous from the right in all points of $(\Omega + \mathbb{Z}_+) \cap (\Omega - \mathbb{Z}_+)$. If $\bar{x}$ is such a discontinuity point then $\lim_{x\downarrow\bar{x}} \hat{Q}(x) - \hat{Q}(\bar{x}) = q^- \Pr\{\omega \in \bar{x} - \mathbb{Z}_+\}$ and $\hat{Q}(\bar{x}) - \lim_{x\uparrow\bar{x}} \hat{Q}(x) = -q^+ \Pr\{\omega \in \bar{x} + \mathbb{Z}_+\}$. In particular this is true for all points in the support Ω.

In between two successive points in $\Omega + \mathbb{Z}$ the function $\hat{Q}$ is constant. In particular, if for some fixed scalar a, $\Omega \subset a + \mathbb{Z}$ (that is, $\omega - a$ is an integer valued random variable), then the function $\hat{Q}$ is constant on all intervals $(a+n, a+n+1)$, $n \in \mathbb{Z}$. Moreover, in that case $\hat{Q}(a+n) = Q(a+n)$ for all such n. □

Exercise 4.2.22 Using part (c) of Theorems 4.2.12 and 4.2.18, show that $\hat{Q}(x) \leq Q(x) + \max\{q^+, q^-\}$, $x \in \mathbb{R}$. ◁

Exercise 4.2.23 Prove the bound on the Lipschitz constant L given in part (e) of Theorem 4.2.21. Hint: Use that $L = \max\{L_+, L_-\}$, where $L_+ = \sup_{x\in\mathbb{R}} |\hat{Q}'_+(x)|$ and $L_- = \sup_{x\in\mathbb{R}} |\hat{Q}'_-(x)|$. ◁

The following examples illustrate various parts of Theorem 4.2.21.

Example 4.2.24 Assume that ω follows an exponential distribution with parameter λ. Using (4.10) in Theorem 4.2.21 (a) and using Examples 4.2.14 and 4.2.19, it can be shown that

$$\hat{Q}(x) = q^+ \left(\lceil -x \rceil^+ + \frac{e^{-\lambda(x + \lceil -x \rceil^+)}}{1 - e^{-\lambda}} \right) + q^- \left(\lceil x \rceil^+ - e^{-\lambda x} \frac{1 - e^{\lambda \lceil x \rceil^+}}{1 - e^{\lambda}} \right), \quad x \in \mathbb{R}.$$

See Fig. 4.8 for the graph of this function $\hat{Q}$.

In line with Theorem 4.2.21 (d), the function $\hat{Q}$ is semi-periodic on $(-\infty, 0]$ with slope $-q^+ = -1$, since $\Pr\{\omega < 0\} = 0$. Moreover, in line with Theorem 4.2.21 (e), the function $\hat{Q}$ is Lipschitz continuous with constant $L \leq \max\{q^+, q^-\}(1 + |\Delta| f/2) = 2(1 + f(0)) = 2 + 2\lambda$. Furthermore, since the exponential density function f is continuous except at $x = 0$ it follows that $\hat{Q}$ is differentiable at all

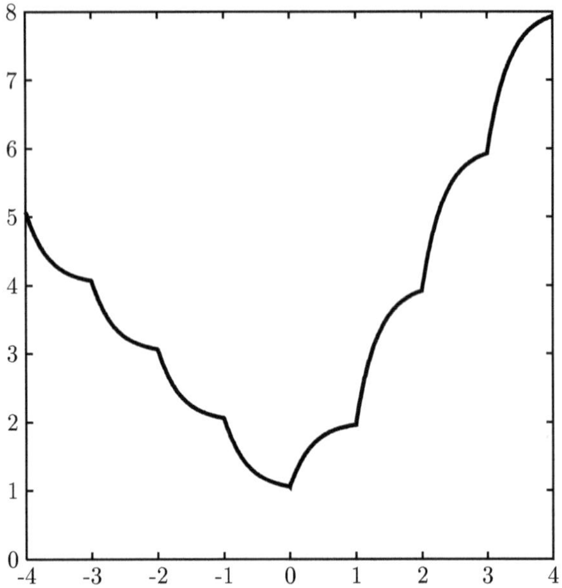

Fig. 4.8 The function $\hat{Q}$ if $\omega \sim \mathcal{E}(3)$, and $q^+ = 1$, $q^- = 2$

$x \notin \mathbb{Z}$ but not at $x \in \mathbb{Z}$. Indeed, these points $x \in \mathbb{Z}$ correspond to the kink points in the graph of $\hat{Q}$ in Fig. 4.8. ◇

Example 4.2.25 Assume that ω is discretely distributed with $\Pr\{\omega = 1\} = 0.25$ and $\Pr\{\omega = 5.3\} = 0.75$, and that $q^+ = 1$, $q^- = 1.5$, see Fig. 4.9. Then $\hat{Q}$ is lower semicontinuous and discontinuous at all points $\{1, 5.3\} + \mathbb{Z}$. In particular, in line with Theorem 4.2.21 (g),

(i) $\hat{Q}$ is continuous from the left but not from the right at all points $\{1, 5.3\} + \{1, 2, \dots\}$. If $\bar{x} \in 1 + \{1, 2, \dots\}$ then the size of the discontinuity at $\bar{x}$ is $q^- \Pr\{\omega \in \bar{x} - \mathbb{Z}_+\} = 1.5 \Pr\{\omega = 1\} = 0.375$. Similarly, if $\bar{x} \in 5.3 + \{1, 2, \dots\}$ then the size of the discontinuity at $\bar{x}$ is $1.5 \Pr\{\omega = 5.3\} = 1.125$.
(ii) $\hat{Q}$ is continuous from the right but not from the left at all points $\{1, 5.3\} - \{1, 2, \dots\}$. If $\bar{x} \in 1 - \{1, 2, \dots\}$ then the size of the discontinuity at $\bar{x}$ is $-q^+ \Pr\{\omega \in \bar{x} + \mathbb{Z}_+\} = -\Pr\{\omega = 1\} = -0.25$. Similarly, if $\bar{x} \in 5.3 - \{1, 2, \dots\}$ then the size of the discontinuity at $\bar{x}$ is $-\Pr\{\omega = 5.3\} = -0.75$.
(iii) $\hat{Q}$ is neither continuous from the left nor continuous from the right at $\bar{x} = 1$ and $\bar{x} = 5.3$. At $\bar{x} = 1$, the size of the left and right discontinuities are -0.25 and 0.375, respectively, whereas these jumps are -0.75 and 1.125 at $\bar{x} = 5.3$.

Moreover, $\hat{Q}$ is semi-periodic with period 1 on the intervals $(-\infty, 1]$, $[1, 5.3]$ and $[5.3, \infty)$, with slopes $-q^+ = -1$, $-q^+ \Pr\{\omega > 1\} + q^- \Pr\{\omega < 5.3\} = -0.375$ and $q^- = 1.5$, respectively. ◇

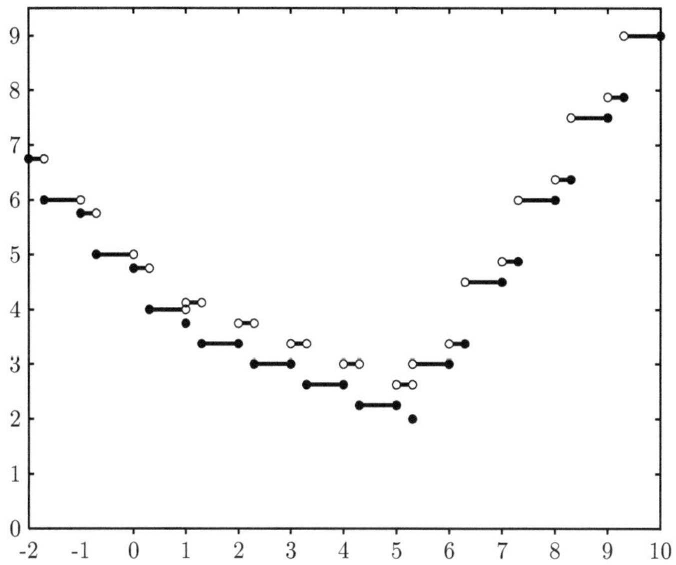

Fig. 4.9 The function $\hat{Q}$ if ω is discretely distributed with $\Pr\{\omega = 1\} = 0.25$ and $\Pr\{\omega = 5.3\} = 0.75$, and $q^+ = 1, q^- = 1.5$

4.3 Algorithms for Mixed-Integer Recourse Models

We now discuss solution methods for mixed integer recourse problems. Contrary to the case with only continuous decision variables, even from a theoretical point of view these problems are not nice. In particular, they are *non-convex*, so that even if we could evaluate Q (which we can not in general) the problem is still *difficult to solve*. Not surprisingly, most solution methods that are available apply to special cases, in particular to the simple integer recourse problem and to problems with discrete distributions.

First we survey algorithms for simple integer recourse models that are all based on convex approximations. Next we review algorithms for general mixed-integer recourse models, which are based on ideas originating from stochastic programming and/or (mixed-)integer programming.

4.3.1 Simple Integer Recourse

Algorithms for the simple integer recourse problem, that was introduced in Sect. 4.2.2, benefit from the additional structure in the problem. The algorithms apply to models with randomness only in the right-hand side parameter ω, and as we have seen this makes the problem separable in the tender variables Tx. This allows to present the results in a one-dimensional setting.

Disregarding problems caused by integer first-stage variables, the difficulty is that the (one-dimensional) integer expected recourse function $\hat{Q}$ is non-convex in general. Therefore, the approach to solving these problems has been to find good convex approximations of $\hat{Q}$, which would allow to solve the problem at least approximately by convex optimization algorithms.

An obvious candidate for such a convex approximation of $\hat{Q}$ is the corresponding continuous simple recourse function Q. However, by Theorem 4.2.21 we know that $Q(x) \leq \hat{Q}(x) \leq Q(x) + \max\{q^+, q^-\}$, $x \in \mathbb{R}$, so that depending on the values of q^+ and q^- this may be a very poor approximation.

Since the function $\hat{Q}$ is convex on any grid $\alpha + \mathbb{Z}$, $\alpha \in [0, 1)$, a convex approximating model is obtained if $\hat{Q}$ is replaced by the piecewise linear function generated by the function values on such a grid. This approach seems attractive from a computational point of view, but since it is an approximation it requires an error bound to guarantee the performance of the resulting approximating solution. Below in Theorem 4.3.7 we will see that such a bound can be obtained. At the end of this section it is shown that this bound can be improved.

Exercise 4.3.1 Show that this approximation yields exact solutions if either the tender variable is restricted to be integer or if ω is distributed on a subset of $\alpha + \mathbb{Z}$ for some $\alpha \in [0, 1)$. ◁

Given the trivial convex bounds Q and $Q + \max\{q^+, q^-\}$, it is reasonable to require that any non-trivial convex approximation of $\hat{Q}$ should be between these bounds too. For such approximations, the following result is the key to solving the corresponding approximating problem.

Theorem 4.3.2 *Let $\hat{Q}^c$ be a convex function on $\mathbb{R}$ bounded from below by Q and bounded from above by $Q + \max\{q^+, q^-\}$. Then*

$$W(s) = \frac{(\hat{Q}^c)'_+(s) + q^+}{q^+ + q^-}, \qquad s \in \mathbb{R},$$

is a cdf, and for all $x \in \mathbb{R}$

$$\begin{aligned} \hat{Q}^c(x) &= q^+ \int_x^\infty (1 - W(s))\, ds + q^- \int_{-\infty}^x W(s)\, ds + \bar{c} \\ &= q^+ \mathbb{E}_\psi \left[(\psi - x)^+\right] + q^- \mathbb{E}_\psi \left[(x - \psi)^+\right] + \bar{c}. \end{aligned}$$

Here, ψ is any random variable with cdf W and mean value $\nu := \mu + \frac{c_1 - c_2}{q^+ + q^-}$, and

$$\bar{c} = \frac{q^+ c_2 + q^- c_1}{q^+ + q^-},$$

where $c_1 := \lim_{x\to-\infty}\{\hat{Q}^c(x) - Q(x)\}$ *and* $c_2 := \lim_{x\to\infty}\{\hat{Q}^c(x) - Q(x)\}$ *satisfying* $0 \le c_i \le \max\{q^+, q^-\}$. □

Theorem 4.3.2 states that every convex function $\hat{Q}^c$ between Q and $Q + \max\{q^+, q^-\}$ is equal to the one-dimensional expected value function of some continuous simple recourse program (plus a constant). In other words, *every reasonable convex approximation of an integer simple recourse problem is equivalent to some continuous simple recourse problem.* Consequently, we can solve simple integer recourse problems (at least approximately) by algorithms developed for continuous simple recourse problems.

However, Theorem 4.3.2 is non-constructive in the sense that the distribution of the random variable ψ is a transformation of the right derivative of $\hat{Q}^c$, which is not known. In the remainder of this section we consider techniques for constructing particular convex approximations $\hat{Q}^c$ or, equivalently, for finding the distribution of ψ that replaces ω in the equivalent continuous simple recourse model.

The Convex Hull

First we assume that ω is a *discrete* random variable with *finite support* $\Omega = \{\omega^1, \dots, \omega^S\}$, and also that the mass points are ordered, that is, $\omega^1 < \omega^2 < \cdots < \omega^S$. In this case it is possible to find the *convex hull* of $\hat{Q}$. (See Definition B.31 and the accompanying discussion for properties of the convex hull.)

By Theorem 4.2.21, under these assumptions the function $\hat{Q}$ has the following properties:

(a) $\hat{Q}$ is a finite function, that is discontinuous at all points of $\mathcal{D}_\infty$, where

$$\mathcal{D}_\infty = \bigcup_{s=1}^{S}\{\omega^s + \mathbb{Z}\}.$$

(b) $\hat{Q}$ is constant in between successive discontinuity points;
(c) $\hat{Q}$ is lower semicontinuous;
(d) $\hat{Q}$ is semi-periodic with period 1 on the intervals $(-\infty, \omega^1]$ and $[\omega^S, \infty)$, with slope $-q^+$ and q^-, respectively.

The crucial observation to make is that, due to properties (a)–(c) above, the convex hull of the function $\hat{Q}$ depends simply and solely on the points $(d, \hat{Q}(d))$, $d \in \mathcal{D}_\infty$. It follows that the convex hull of $\hat{Q}$ is equal to the convex hull of the piecewise linear function $\hat{Q}^{pl}$ generated by these points.

The convex hull of a piecewise linear function f can be determined by an algorithm known as the *Graham scan* [9], which is based on the observation that a knot $(d, f(d))$ can be on the convex hull only if $f'_-(d) \le f'_+(d)$, where f_- and f_+ are the left and right derivative of f, respectively. Thus, if this condition is not satisfied then we may eliminate the knot d^i, and redefine the function f

on $[d^{i-1}, d^{i+1}]$ as the linear function connecting the points $(d^{i-1}, f(d^{i-1}))$ and $(d^{i+1}, f(d^{i+1}))$. It is easy to see that by repeating this procedure until all knots satisfy the condition mentioned above, eventually we end up with the convex hull of the function f.

Of course, this algorithm only terminates in finite time if only a finite number of knots has to be to considered. For the function $\hat{Q}^{pl}$ finiteness of the initial set of knots is due to its semi-periodicity on the intervals $(-\infty, \omega^1]$ and $[\omega^S, \infty)$ (property (d) above), which implies that that conv $\hat{Q}$ is affine on $(-\infty, \omega^1-1]$ and $[\omega^S+1, \infty)$ with slopes $-q^+$ and q^-, respectively. (Verify this.) Consequently, we only need to determine the convex hull of $\hat{Q}^{pl}$ restricted to $[\omega^1 - 1, \omega^S + 1]$ which depends only on knots in the set $\mathcal{D} = \mathcal{D}_\infty \cap [\omega^1 - 1, \omega^S + 1]$, which is finite indeed. The convex hull of $\hat{Q}^{pl}$ restricted to $[\omega^1 - 1, \omega^S + 1]$ can be extended to its convex hull on $\mathbb{R}$ in a trivial way.

Exercise 4.3.3 This procedure only uses one-sided derivatives, so that we only need to know the difference between function values in neighboring discontinuity points of $\hat{Q}$. Let $d^1 < d^2$ be any two such points. Show that

$$\begin{aligned}\hat{Q}(d^2) - \hat{Q}(d^1) = -q^+ \sum_{s=1}^{S} &\left\{\Pr\left\{\omega = \omega^s\right\} : d^2 \in \omega^s - \mathbb{Z}_+\right\} \\ +q^- \sum_{s=1}^{S} &\left\{\Pr\left\{\omega = \omega^s\right\} : d^1 \in \omega^s + \mathbb{Z}_+\right\}.\end{aligned}$$

Thus, for each point $d \in \mathcal{D}$ we simply have to record by which mass point(s) it is generated. ◁

After completing the procedure outlined above, we know all knots of conv $\hat{Q}$ and its slope in between two successive knots. That is, conv $\hat{Q}$ is determined up to a constant. By Theorem 4.3.2 this information is sufficient to determine the distribution of the random variable ψ that replaces ω in the equivalent continuous simple recourse formulation. Indeed, it follows that ψ has a cdf that is a simple transformation of the known right derivative of conv $\hat{Q}$. Since conv $\hat{Q}$ is piecewise linear, we see that *ψ is a discrete random variable with mass points ψ^i that correspond to the knots of* conv $\hat{Q}$. Moreover, $\Pr\left\{\psi = \psi^i\right\} = \delta^i/(q^+ + q^-)$, where δ^i is the increase of the slope of conv $\hat{Q}$ at ψ^i. It remains to compute the constant $\bar{c}$ as defined in Theorem 4.3.2. We use that conv $\hat{Q}$ equals $\hat{Q}$ at each of its knots. Hence,

$$\bar{c} = \hat{Q}(\psi^i) - \left(q^+ \mathbb{E}_\psi\left[(\psi - \psi^i)^+\right] + q^- \mathbb{E}_\psi\left[(\psi - \psi^i)^-\right]\right),$$

where ψ^i is an arbitrary point in the support of ψ. This is the only time that we actually calculate a function value of $\hat{Q}$.

Remark 4.3.4 See [18] for details of this algorithm, including a speedup based on semi-periodicity of $\hat{Q}$ on each interval $[\omega^s, \omega^{s+1}]$, $s = 1, \dots, S-1$.

Thus, we can compute the convex hull of the one-dimensional function $\hat{Q}$ efficiently. However, we actually need to find the *convex hull of the n-dimensional expected value function Q*, given by

$$Q(x) = \sum_{i=1}^{m} \hat{Q}_i(T_i x), \quad x \in \mathbb{R}^n,$$

with

$$\hat{Q}_i(T_i x) = q_i^+ \mathbb{E}_{\omega_i}\left[\lceil \omega_i - T_i x \rceil^+\right] + q_i^- \mathbb{E}_{\omega_i}\left[\lfloor \omega_i - T_i x \rfloor^-\right],$$

where the scalars q_i^+ and q_i^- denote the ith components of vectors q^+ and q^-, respectively, ω_i denotes the ith component of the random vector $\omega = (\omega_1, \dots, \omega_m)$, and T_i is the ith row of the matrix T. In [17] it is shown that, at least if the matrix T has linear independent rows,

$$\operatorname{conv} Q(x) = \sum_{i=1}^{m} \operatorname{conv} \hat{Q}_i(T_i x), \quad x \in \mathbb{R}^n.$$

That is, the convex hull of Q is determined by the convex hulls of the functions $\hat{Q}_i$ in precisely the same way as Q itself is determined by the functions $\hat{Q}_i$.

Remark 4.3.5 In general it is *not* true that the convex hull of a sum of functions is equal to the sum of the convex hulls of the individual functions. Consider for example

$$f_1(x) = \begin{cases} x^2, & \text{if } x \in (-\infty, 0) \cup (0, \infty); \\ 1, & \text{if } x = 0, \end{cases}$$

$$f_2(x) = \begin{cases} 0, & \text{if } x \in (-\infty, 0) \cup (0, \infty); \\ -1, & \text{if } x = 0, \end{cases}$$

and $f = f_1 + f_2$. It is easy to see that $\operatorname{conv} f(x) = f(x) = x^2$, whereas $\operatorname{conv} f_1(x) = x^2$ and $\operatorname{conv} f_2(x) = -1$, so that $\operatorname{conv} f_1(x) + \operatorname{conv} f_2(x) = x^2 - 1$.

Also, it is *not* true in general, that the convex hull of a composition of a function f with a linear transformation is equal to the composition of the convex hull of f with that linear transformation. Consider

$$f(x_1, x_2) = e^{x_1^2 - x_2^2}, \quad (x_1, x_2) \in \mathbb{R}^2,$$

$$f_S(y) = f(Sy), \quad y \in \mathbb{R}, \; S = \begin{bmatrix} 1 \\ 1 \end{bmatrix}.$$

Then $\text{conv} f(x_1, x_2) = 0$ for all $(x_1, x_2) \in \mathbb{R}^2$, so that $\text{conv} f(Sy) = 0$ for all $y \in \mathbb{R}$, whereas $\text{conv} f_S(y) = f_S(y) = 1$ for all $y \in \mathbb{R}$.

Finally, it should be noted that the above calculation of the convex hull does not depend on the first-stage constraints $Ax = b$ and the non-negativities $x \geq 0$. This provides no hardship if the constrained optimum coincides with the free optimum. However, if this is not the case then it may happen that

$$\inf\{cx + Q(x) : Ax = b,\ x \geq 0\} > \inf\{cx + \text{conv}\, Q(x) : Ax = b,\ x \geq 0\},$$

so that we only obtain a lower bound on the optimal value.

Convexity by Perturbation of the Distribution[1]

In this section we discuss a particular class of convex approximations of the one-dimensional function $\hat{Q}$. The basic idea is to approximate the true distribution of the random variable ω by a distribution that, according to Theorem 4.2.21 (f), causes convexity of $\hat{Q}$. The resulting function, say $\hat{Q}^c$, is then a convex approximation of the function $\hat{Q}$.

Let $\mathcal{C}$ denote the class of regular probability density functions satisfying the conditions of Theorem 4.2.21 (d). In principle, we can use any pdf in $\mathcal{C}$ for our approximation, and ideally we would like to find a pdf in $\mathcal{C}$ that is optimal in some sense. For example, a suitable candidate would be one that minimizes the error in the optimal value obtained. However, it is not clear at all how such an optimal pdf can be found. Below we discuss a subclass of $\mathcal{C}$, for which a bound on the approximation error is known, at least if the original *distribution of ω is continuous*.

Starting point for the convex approximations that we have in mind is the observation that the function $\hat{Q}$ is convex if ω is uniformly distributed on the interval $[a, a + 1]$ for some $a \in \mathbb{R}$.

Exercise 4.3.6 Use Theorem 4.2.21 (d) to prove this. ◁

It follows that $\hat{Q}$ is also convex if ω has a pdf that is constant on every interval $[\alpha + k, \alpha + k + 1)$, $k \in \mathbb{Z}$, for some $\alpha \in [0, 1)$. Below we analyze convex approximations of $\hat{Q}$ based on this class of piecewise constant densities. We start by defining how such approximating densities can be constructed for an arbitrary distribution.

Let ω be a random variable with cdf F. For α a shift parameter in $[0, 1)$, define the continuous random variable ω_α with pdf f_α,

$$f_\alpha(s) = F(\lfloor s \rfloor_\alpha + 1) - F(\lfloor s \rfloor_\alpha), \quad s \in \mathbb{R},$$

[1] See [19] for details and generalizations of results in this section. Corresponding results on convex approximations for complete integer recourse models are described in [36, 48] and generalized in [37].

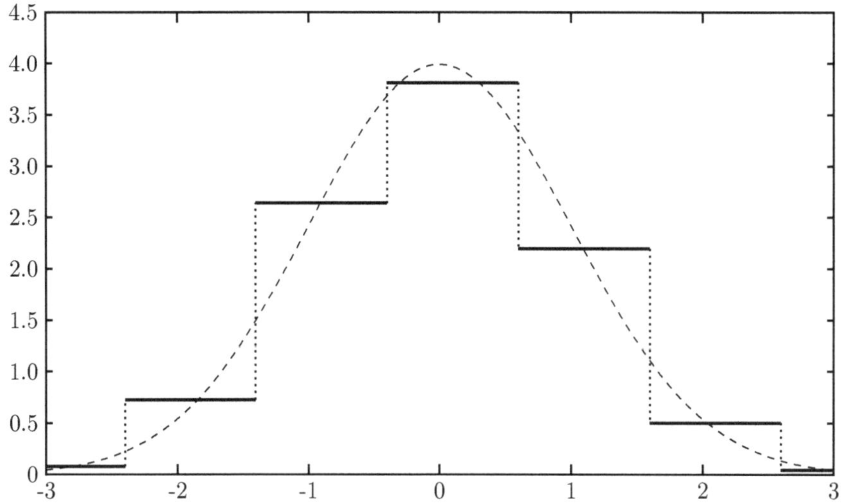

Fig. 4.10 The pdf f (dashed) and f_α (solid) with $\alpha = 0.6$, for the standard normal distribution

where $\lfloor s \rfloor_\alpha := \max\{\alpha + k : \alpha + k \leq s,\ k \in \mathbb{Z}\}$ denotes round down with respect to the lattice $\alpha + \mathbb{Z}$. For example, $\alpha = 0$ corresponds to the usual integer round down and $\lfloor 2.1 \rfloor_{0.9} = 1.9$. Similarly, round up with respect to $\alpha + \mathbb{Z}$ is defined as $\lceil s \rceil_\alpha := \min\{\alpha + k : \alpha + k \geq s,\ k \in \mathbb{Z}\}$.

From its definition, we see that f_α is indeed constant on every unit interval $[\alpha + k, \alpha + k + 1)$, $k \in \mathbb{Z}$. The distribution of ω_α is related to the distribution of ω in the following way: to every interval $[\alpha + k, \alpha + k + 1)$, the pdf f_α of ω_α assigns the same probability mass as the distribution F of ω. See Fig. 4.10.

The random variable ω_α is called the *α-approximation* of ω. It is used to construct convex approximations of the function $\hat{Q}$ as defined below.

Theorem 4.3.7 *Let ω be a continuous random variable with pdf f. For all $\alpha \in [0, 1)$, let ω_α be an α-approximation of ω. For each $\alpha \in [0, 1)$ define* the α-approximation of $\hat{Q}$ *as*

$$\hat{Q}_\alpha(x) = q^+ \mathbb{E}_{\omega_\alpha}\left[\lceil \omega_\alpha - x \rceil^+\right] + q^- \mathbb{E}_{\omega_\alpha}\left[\lfloor \omega_\alpha - x \rfloor^-\right], \quad x \in \mathbb{R},$$

where q^+ and q^- are non-negative scalars. Then

(a) For each α, the function $\hat{Q}_\alpha$ is convex.

It is related to the function $\hat{Q}(x) = q^+ \mathbb{E}_\omega\left[\lceil \omega - x \rceil^+\right] + q^- \mathbb{E}_\omega\left[\lfloor \omega - x \rfloor^-\right]$ by the equation

$$\hat{Q}_\alpha(x) = \hat{Q}(\lfloor x \rfloor_\alpha) + (x - \lfloor x \rfloor_\alpha)\left(\hat{Q}(\lceil x \rceil_\alpha) - \hat{Q}(\lfloor x \rfloor_\alpha)\right),$$

so that $\hat{Q}_\alpha$ *is the piecewise linear function that coincides with* $\hat{Q}$ *at all points* $\alpha + k$, $k \in \mathbb{Z}$.

(b) Assume that the pdf $f \in \mathcal{R}$ *is regular (see Sect. A.3). Then, for all* $x \in \mathbb{R}$,

$$|\hat{Q}_\alpha(x) - \hat{Q}(x)| \leq \min\left\{x - \lfloor x \rfloor_\alpha, \lceil x \rceil_\alpha - x\right\}(q^+ + q^-)\frac{|\Delta| f}{2},$$

so that

$$\|\hat{Q}_\alpha - \hat{Q}\|_\infty \leq (q^+ + q^-)\frac{|\Delta| f}{4}.$$

In particular, if the pdf f *is unimodal then*

$$\|\hat{Q}_\alpha - \hat{Q}\|_\infty \leq (q^+ + q^-)\frac{f(\nu)}{2},$$

where ν *is the mode of the distribution.* □

We see that $\hat{Q}_\alpha$ is precisely the piecewise linear function generated by the restriction of $\hat{Q}$ to $\{\alpha + \mathbb{Z}\}$, which we already know to be convex by Theorem 4.2.21. However, due to its more tractable representation, it is now possible to give a bound on the error of the approximation as specified in part (b) of the theorem.

The error bound is proportional to the total variation $|\Delta| f$ of the pdf f of ω, which equals $2f(\nu)$ for unimodal distributions with node ν. For such distributions (e.g. normal and exponential) it holds that $|\Delta| f$ decreases as the variance increases, so that the error bound implies that $\hat{Q}_\alpha$ is a better approximation of $\hat{Q}$ as ω has a higher variance. For such distributions $\hat{Q}$ is almost convex if the variance of ω is sufficiently large, in the sense that it is very close (in the supremum norm) to any of the convex functions $\hat{Q}_\alpha$, $\alpha \in [0, 1)$.

Next we give the equivalent continuous simple recourse formulation of the convex approximation $\hat{Q}_\alpha$.

Theorem 4.3.8 *Let* ω *be a continuous random variable with cdf* F *and* $\alpha \in [0, 1)$. *Denote, as before, the* α*-approximations of the one-dimensional expected value function by* $\hat{Q}_\alpha$. *Then, for* $x \in \mathbb{R}$,

$$\hat{Q}_\alpha(x) = q^+\mathbb{E}_{\psi_\alpha}\left[(\psi_\alpha - x)^+\right] + q^-\mathbb{E}_{\psi_\alpha}\left[(\psi_\alpha - x)^-\right] + \frac{q^+q^-}{q^+ + q^-}, \quad (4.11)$$

where ψ_α *is a random variable with cdf*

$$W_\alpha(s) = \frac{q^+}{q^+ + q^-}F(\lfloor s \rfloor_\alpha) + \frac{q^-}{q^+ + q^-}F(\lfloor s \rfloor_\alpha + 1), \quad s \in \mathbb{R}.$$

That is, ψ_α is a discrete random variable with support $\Psi_\alpha \subset \{\alpha + \mathbb{Z}\}$ and

$$\Pr\{\psi_\alpha = \alpha + k\} = \frac{q^+}{q^+ + q^-}\big(F(\alpha + k) - F(\alpha + k - 1)\big)$$
$$+ \frac{q^-}{q^+ + q^-}\big(F(\alpha + k + 1) - F(\alpha + k)\big), \quad k \in \mathbb{Z}.$$

□

Note that the distribution of ψ_α can be computed directly from the distribution of ω, and that the constant in (4.11) does not depend on α.

Example 4.3.9 Assume that ω follows an exponential distribution with parameter λ (notation: $\omega \sim \mathcal{E}(\lambda)$), and that $q^+ = q^- = 1$. Then, for every $\alpha \in [0, 1)$,

$$\|\hat{Q}_\alpha - \hat{Q}\|_\infty \leq (q^+ + q^-)\frac{|\Delta| f}{4} = f(0) = \lambda,$$

where we used that the pdf f of the exponential distribution is unimodal with mode 0.

Using Theorem 4.3.8 it follows that the random variable ψ_α, that replaces ω in the continuous simple recourse representation of $\hat{Q}_\alpha$, is discretely distributed on $\{\alpha + k\}$, $k = -1, 0, 1, \ldots$, with

$$\Pr\{\psi_\alpha = \alpha + k\} = \begin{cases} \dfrac{1 - e^{-\lambda(\alpha+k+1)}}{2}, & k \in \{-1, 0\} \\ \dfrac{e^{-\lambda(\alpha+k)}\left(e^{\lambda} - e^{-\lambda}\right)}{2}, & k \geq 1. \end{cases}$$

See Fig. 4.11 for the plots of $\hat{Q}$, $\hat{Q}_\alpha$, and the distribution of ψ_α. ◇

Improved Error Bound

Theorem 4.3.7 provides an error bound for the α-approximation $\hat{Q}_\alpha$. Romeijnders et al. [37] derive improved error bounds. In this section we will illustrate the derivation of this improved error bound for the integer expected surplus function $\hat{G}$ and its α-approximation

$$\hat{G}_\alpha(x) = \mathbb{E}_\omega\left[(\lceil \omega - \alpha \rceil + \alpha - x)^+\right], \quad x \in \mathbb{R}.$$

Here, $\hat{G}_\alpha$ is a one-dimensional continuous surplus function with ψ_α, as defined in Theorem 4.3.8, given by $\psi_\alpha = \lceil \omega - \alpha \rceil + \alpha$, since $q^+ = 1$ and $q^- = 0$. Observe that $\hat{G}_\alpha$ is the piecewise linear function generated by the restriction of $\hat{G}$ to $\{\alpha + \mathbb{Z}\}$. Indeed, for $x \in \{\alpha + \mathbb{Z}\}$, we have $\hat{G}_\alpha(x) = \hat{G}(x)$.

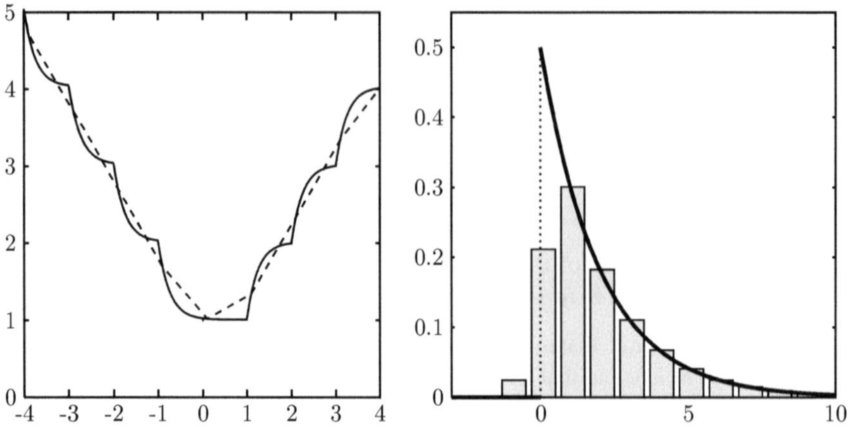

Fig. 4.11 Illustration of Example 4.3.9 with $q^+ = q^- = 1$ and $\alpha = 0.1$. Left: The functions $\hat{Q}$ and $\hat{Q}_\alpha$ if $\omega \sim \mathcal{E}(5)$. Right: The pdf of $\omega \sim \mathcal{E}(0.5)$ and the corresponding discrete distribution of ψ_α (denoted by bars centered at $\alpha + k$, $k = -1, 0, 1, \dots$)

To derive an upper bound on $|\hat{G}(x) - \hat{G}_\alpha(x)|$, $x \in \mathbb{R}$, consider the underlying difference function $\psi_{\alpha,x}$ defined for fixed $\alpha \in \mathbb{R}$ and $x \in \mathbb{R}$ as

$$\psi_{\alpha,x}(t) = \lceil t - x \rceil^+ - (\lceil t - \alpha \rceil + \alpha - x)^+, \quad t \in \mathbb{R}.$$

For this difference function, $\hat{G}(x) - \hat{G}_\alpha(x) = \mathbb{E}_\omega \left[\psi_{\alpha,x}(\omega) \right]$, $x \in \mathbb{R}$. If we ignore the positive part operators in the expression of $\psi_{\alpha,x}$ for the moment, then we obtain $\varphi_{\alpha,x}$ given by

$$\varphi_{\alpha,x}(t) = \Big(\lceil t - x \rceil + x \Big) - \Big(\lceil t - \alpha \rceil + \alpha \Big), \quad t \in \mathbb{R}.$$

This function $\varphi_{\alpha,x}$ has several special properties. First of all, $\varphi_{\alpha,x}$ is periodic in t with period 1. Second, it is symmetric in α and x in the sense that $\varphi_{\alpha,x} = -\varphi_{x,\alpha}$ for all $\alpha \in \mathbb{R}$ and $x \in \mathbb{R}$, and third it has zero average:

$$\int_0^1 \varphi_{\alpha,x}(t)dt = 0.$$

It turns out that a worst-case analysis of $\mathbb{E}_\omega \left[\varphi_{\alpha,x}(\omega) \right]$ is useful. We consider for every periodic function φ,

$$M(\varphi, B) := \sup_{f \in \mathcal{F}} \Big\{ \mathbb{E}_f \left[\varphi(\omega) \right] : |\Delta| f \le B \Big\}, \quad B > 0.$$

That is, we consider the worst-case expected value $\mathbb{E}_f[\varphi(\omega)]$ over all regular probability density functions $f \in \mathcal{R}$ (see Sect. A.3) with total variation $|\Delta| f \leq B$. For special case of periodic functions the value of $M(\varphi, B)$ can be determined exactly. In fact, for every $\alpha \in \mathbb{R}$ and $x \in \mathbb{R}$,

$$M(\varphi_{\alpha,x}, B) = \min\{\gamma_{\alpha,x}, \gamma_{\alpha,x}(1 - \gamma_{\alpha,x})B/2\},$$

where $\gamma_{\alpha,x} = (x - \alpha) - \lfloor x - \alpha \rfloor$. Maximizing over α and x for every fixed $B > 0$ yields $\sup_{\alpha,x} M(\varphi_{\alpha,x}, B) = h(B)$, where

$$h(B) = \begin{cases} B/8, & 0 < B \leq 4, \\ 1 - 2/B, & B \geq 4. \end{cases}$$

Interestingly, the same bound that applies to $\varphi_{\alpha,x}$ also applies to $\psi_{\alpha,x}$, and thus

$$|\hat{G}(x) - \hat{G}_\alpha(x)| \leq h(|\Delta| f), \quad x \in \mathbb{R}.$$

This is an improvement by a factor 2 over the bound in Theorem 4.3.7 for $0 < B \leq 4$, and an even bigger improvement for $B > 4$.

4.3.2 General Mixed-Integer Recourse

In this section we discuss three algorithms for mixed-integer recourse problems. This selection is representative for the existing algorithms, in that they are inspired by results from continuous stochastic programming and deterministic mixed-integer programming, as all such algorithms are. For a survey of general algorithms we refer to Klein Haneveld and Van der Vlerk [20]; several applications and special purpose algorithms can be found in Stougie and Van der Vlerk [44].

Consider the case that Ω is finite. That is, $\Omega = \{\omega^1, \ldots, \omega^S\} \subset \mathbb{R}^r$, and $\Pr\{\omega = \omega^s\} = p_s$, $s = 1, \ldots, S$ is its known distribution. Then, using the notation $T^s := T(\omega^s)$, $h^s := h(\omega^s)$, MIR can be written as a large scale mixed-integer programming problem:

$$\begin{array}{llllll} \min\limits_{\substack{x\in\mathbb{R}^n \\ y^s\in\mathbb{R}^p \\ s=1,\ldots,S}} & cx + p_1 \cdot qy^1 + p_2 \cdot qy^2 + \ldots + p_S \cdot qy^S & & & & \\ & Ax & & & & \sim b \\ & T^1x + & Wy^1 & & & \sim h^1 \\ & T^2x & & + \quad Wy^2 & & \sim h^2 \\ & \vdots & & & \ddots & \\ & T^Sx & & & + \quad Wy^S & \sim h^S \\ & x \in X & y^1 \in Y & y^2 \in Y & y^S \in Y & \end{array} \tag{4.12}$$

Here $y^s = y(\omega^s)$ indicates the recourse actions to be taken if the realization ω^s of ω occurs. Recall that integrality restrictions are given by the sets $X = \{x \in \mathbb{Z}^{\bar{n}} \times \mathbb{R}^{n-\bar{n}} : x^l \leq x \leq x^u\}$ and $Y = \{y \in \mathbb{Z}^{\bar{p}} \times \mathbb{R}^{p-\bar{p}} : y^l \leq y \leq y^u\}$. For any reasonable size of the underlying problem and / or the number of realizations S, this is indeed a large scale problem: it has $m_1 + Sm$ (explicit) constraints and $n + Sp$ variables, of which $\bar{n} + S\bar{p}$ are required to be integer. Thus, in almost all applications it is impossible to solve this problem without using its special structure, which is precisely what the first two algorithms below aim at.

L-Shaped Methods

In the large-scale formulation of MIR we can consider the decision vector x to be the complicating factor, since if x were fixed the remaining problem would be separable and therefore much easier to solve. For general deterministic mixed-integer problems this approach of fixing complicating variables (which are chosen to be the integer variables, so that the remaining problem is an LP) has led to Benders' decomposition.

Similarly, for continuous two-stage recourse models we may consider the first-stage variables x as 'complicating' the separability. Based on this observation, Van Slyke and Wets [49] proposed the L-shaped algorithm (see Sect. 3.4.2).

In both cases, the part of the objective function corresponding to the remaining variables (i.e., the expected value function in the two-stage model) is represented by a single variable θ, whose optimal value is determined iteratively by using LP duality to construct a piecewise linear outer approximation of this convex part of the objective function as well as its effective domain. The cuts that are generated to these ends are known as optimality cuts and feasibility cuts, respectively.

Given these very similar solution methods for these two problem classes, it is only natural to combine them to obtain an algorithm for mixed-integer recourse problems. In the discussion below we do not consider feasibility cuts, that is, we assume that the problems have (relatively) complete recourse.

This approach was first followed by Wollmer [52] for models with $0 - 1$ first-stage variables and continuous second-stage problems. Since the expected value function Q is convex in this case, this is a straightforward combination of Benders' decomposition and the L-shaped method.

Laporte and Louveaux [23] extend this approach to models with binary first-stage variables and mixed-integer second-stage problems. Of course, in general the expected value function Q of such a problem is non-convex, so that it cannot be described using linear cuts as in the continuous case. However, due to the fact that the first-stage variables are binary, it is possible to construct a valid set of linear optimality cuts for this problem. For $x^k \in X \subset \{0, 1\}^n$ define the index set $S^k = \{i : x_i^k = 1\}$. Then an optimality cut at x^k is given by

$$\theta \geq \left(Q(x^k) - L\right)\left(\sum_{i \in S^k} x_i - \sum_{i \notin S^k} x_i - |S^k| + 1\right) + L, \tag{4.13}$$

where L is a global lower bound for Q.

Exercise 4.3.10 Verify that this cut is sharp at x^k and not larger than $Q(x)$ for all $x \in \{0, 1\}^n$. Note that it may not be valid for $x \in (0, 1)^n$, but such x are not feasible anyway. ◁

Since there can be a large number of these optimality cuts (2^n at most), they are generated iteratively in a branch and cut scheme, which is illustrated in the example below. This so-called integer L-shaped method has been applied successfully to medium sized problems with easily computable second-stage.

Example 4.3.11 Consider the integer recourse problem

$$\min\{x + Q(x) : x \in \{0, 1\}\},$$

where

$$Q(x) = \mathbb{E}_\omega\left[v(x, \omega)\right]$$
$$v(x, \omega) = \min\{1.5y : y \geq \omega - x,\ y \in \mathbb{Z}_+\} = 1.5\lceil \omega - x\rceil^+,$$

and $\Pr\{\omega = 1.3\} = \Pr\{\omega = 2.7\} = 0.5$. An obvious lower bound for Q is $L = 0$.

The integer L-shaped method proceeds by iteratively solving LP problems

$$\begin{aligned}\min\{x + \theta :\ & E_t x + \theta \geq e_t,\ t = 1, \ldots, T \\ & x \in [0, 1],\ \theta \in \mathbb{R}\},\end{aligned}$$

where the optimality cuts $E_t x + \theta \geq e_t$ are added when needed, that is, if the current solution (x^k, θ^k) is such that $Q(x^k) > \theta^k$. The cuts are constructed using (4.13).

- At the first iteration ($k = 1$), θ is ignored in the computation so that the problem to solve is $\min\{x : x \in [0, 1]\}$, which has optimal solution $x^1 = 0$. Computing $Q(x^1) = 3.75$ and $S^1 = \emptyset$, we construct the optimality cut $\theta \geq 3.75 - 3.75x$ (see Fig. 4.12).
- For $k = 2$, the current problem is

$$\begin{aligned}\min\{x + \theta :\ & \theta \geq 3.75 - 3.75x \\ & x \in [0, 1],\ \theta \in \mathbb{R}\},\end{aligned}$$

  which has optimal solution $x^2 = 1$, $\theta^2 = 0$. Since $Q(x^2) = 2.25 > \theta^2$, an optimality cut is added: $S^2 = \{1\}$ gives $\theta \geq 2.25x$.

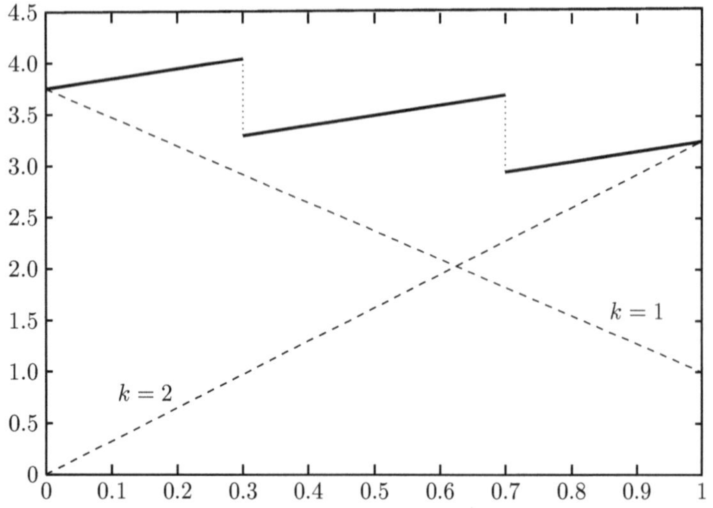

Fig. 4.12 The objective function $cx + Q(x)$ of Example 4.3.11 and its optimality cuts generated by the integer L-shaped method

- For $k = 3$, the current problem is

$$\begin{aligned}\min\{x + \theta :\ & \theta \geq 3.75 - 3.75x \\ & \theta \geq 2.25x \\ & x \in [0, 1],\ \theta \in \mathbb{R}\,\}\end{aligned}$$

  with optimal solution $x^3 = 0.625,\ \theta^3 = 1.40625$. Since $x^3 \notin \{0, 1\}$, a branching step gives two new problems by adding the constraints $x \leq \lfloor x^3 \rfloor$ and $x \geq \lfloor x^3 \rfloor +$ 1, respectively.
- From the preceding calculations we conclude that the optimal solution is $x^\star = 1$, with optimal value 3.25.

◇

The assumption that the first-stage variables are binary is essential in the algorithm of Laporte and Louveaux, since otherwise the linear cuts (4.13) will in general not be valid. If this assumption is dropped, it is necessary to resort to nonlinear optimality cuts for the nonconvex function Q. This approach is worked out by Carøe and Tind [3] for models with arbitrary first-stage and pure integer second-stage variables. Cuts are derived from duality theory for integer programming (see [29]). For a current solution $\bar{x}$ and realizations (T^s, h^s), $s = 1, \ldots, S$, the second-stage problems are

$$\inf_y \{qy : Wy \geq h^s - T^s\bar{x},\ y \in \mathbb{Z}_+^p\},$$

whose respective so-called $\mathcal{F}$-dual problems are

$$\sup_F\{F(h^s - T^s\bar{x}) : F(Wy) \leq qy \; \forall y \in \mathbb{Z}_+^p, \; F \in \mathcal{F}\}, \tag{4.14}$$

where $\mathcal{F}$ is a suitable subset of the set of dual price functions (nondecreasing functions $F : \mathbb{R}^p \mapsto \mathbb{R} \cup \{-\infty, \infty\}$ with $F(0) = 0$). The choice of the set $\mathcal{F}$ is governed by the algorithm that is used to solve the second-stage problems. It should be large enough so that the duality gap is closed.

Theorem 4.3.12 *Suppose that F^s, $s = 1, \ldots, S$, are optimal dual price functions for (4.14). Then an optimality cut for Q at $\bar{x}$ is given by the inequality $\theta \geq \sum_{s=1}^S p^s F^s(h^s - T^s\bar{x})$.* □

The resulting master problems are nonlinear, due to the price functions $F \in \mathcal{F}$ which are in most cases discontinuous, and therefore are in general very hard if not impossible to solve.

Dual Decomposition

The dual decomposition method of Carøe and Schultz [4] applies to two-stage problems with mixed-integer variables in both stages, a complete recourse structure, and a finite discrete distribution $(T, h) = (T^s, h^s)$ with probability p^s, $s = 1, \ldots, S$. The starting point is again the large-scale mixed-integer problem (4.12). Instead of decomposing the problem by time stages, as in the L-shaped methods, the idea is here to decompose the problem by realizations (called scenarios in the multi-stage case). Consequently, the algorithm below is governed by finding good dual multipliers, which is reflected in its name.

We have already seen that the first-stage variables x are complicating variables in the sense that they spoil separability of (4.12). As a first step to remove this obstacle, consider introducing copies x^s, $s = 1, \ldots, S$, and constraints to assure that all copies are equal:

$$\min\left\{\sum_{s=1}^S p^s(cx^s + qy^s) : \begin{array}{l}(x^s, y^s) \in Z^s, \; s = 1\ldots, S\\ x^1 = x^2 = \cdots = x^S\end{array}\right\},$$

where $Z^s = \{(x, y) \in X \times Y : Ax \sim b, Wy + T^s x \sim h^s\}$. Thus, instead of complicating variables x we now have the complicating constraints $x^1 = x^2 = \cdots = x^S$. These constraints (known as non-anticipativity constraints) can be written as $\sum_s H^s x^s = 0$ with $H = (H^1, \ldots, H^S)$ a suitable $(l \times nS)$ matrix. Next, we construct the Lagrangian relaxation with respect to these constraints, which is separable according to realizations (T^s, h^s):

$$D(\lambda) = \sum_{s=1}^S D^s(\lambda), \quad \lambda \in \mathbb{R}^l,$$

where

$$D^s(\lambda) = \min_{x^s, y^s} \left\{ p^s(cx^s + qy^s) + \lambda(H^s x^s) : (x^s, y^s) \in Z^s \right\}, \quad s = 1, \ldots, S.$$

By a well-known weak duality result (see e.g [29]) the optimal value of the Lagrangian dual $z_{LD} := \max_\lambda D(\lambda)$ provides a lower bound for the optimal value of the MIR problem (4.12). Moreover, if for some λ the solutions (x^s, y^s) are feasible then they are optimal, and so is λ.

Thus, instead of directly solving the large-scale mixed-integer formulation of MIR, which is in general impossible because of its dimension $(m_1 + Sm) \times (n + Sp)$, we need to find optimal Lagrangian multipliers λ. The Lagrangian dual is a convex non-smooth problem which can be solved by subgradient methods.

Exercise 4.3.13 Show that the Lagrangian dual $\max_\lambda D(\lambda)$ is a convex problem. Show that $\sum_s H^s x^s$ is a subgradient for D at λ if (x^s, y^s) solve the subproblems $D^s(\lambda)$, $s = 1, \ldots, S$. ◁

At every iteration of such a subgradient method we need to solve S subproblems $D^s(\lambda)$ for a given λ. That is, we need to solve many mixed-integer problems of size $(m_1 + m) \times (n + p)$, which is much smaller than the size of the original problem. It is assumed that the subproblems can be solved quickly.

In general the scenario solutions (x^r, y^r) of the Lagrangian relaxation will not coincide in their x-component. A natural solution to this problem is to take the weighted average $\bar{x} = \sum_s p^s x^s$ as the candidate first-stage solution. However, this solution is in general not feasible due to the integer restrictions on x. Therefore, an additional rounding heuristic is needed to come up with feasible first-stage solutions. In the dual decomposition method this is implemented by means of a branch and bound scheme, which uses the Lagrangian relaxation as a bounding procedure.

Let $\mathcal{P}$ be the collection of remaining problems in the branch and bound tree, and let z_i be the associated value for each problem $P_i \in \mathcal{P}$. Moreover, let $\bar{z}$ be the value of the best feasible solution obtained so far. An iteration of the algorithm consists of the following steps:

(i) Select a current problem P from $\mathcal{P}$, and solve its Lagrangian dual. If the value $z_{LD} \geq \bar{z}$ delete P and proceed with the next iteration.
(ii) If the scenario solutions x^r do not agree, compute the average $\bar{x}$ and apply some rounding heuristic to obtain a feasible first-stage solution, say $\hat{x}$. Otherwise, let $\hat{x} = x^1$.
(iii) If $c\hat{x} + Q(\hat{x}) < \bar{z}$, then update $\bar{z}$ and delete all problems P_i with $z_i \geq \bar{z}$ from the list $\mathcal{P}$.
(iv) Create two new problems by branching on an arbitrary component x_j of x. That is, add constraints $x_j \leq \lfloor \bar{x}_j \rfloor$ and $x_j \geq \lfloor \bar{x}_j \rfloor + 1$ if x_j is an integer component, or $x_j \leq \bar{x}_j$ and $x_j \geq \bar{x}_j$ if not.

The dual decomposition algorithm is applied successfully to problems of realistic size.

Example 4.3.14 Consider the integer recourse problem

$$(P_0) \qquad \min\{x + Q(x) : x \in \mathbb{Z}_+\},$$

where

$$Q(x) = \mathbb{E}_\omega [v(x, \omega)]$$
$$v(x, \omega) = \min\{1.5y : y \ge \omega - x,\ y \in \mathbb{Z}_+\} = 1.5\lceil \omega - x\rceil^+,$$

and $\Pr\{\omega = 1.3\} = \Pr\{\omega = 2.7\} = 0.5$.

The Lagrangian subproblems are, for $s = 1, 2$,

$$D^s(\lambda) = \min\left\{p^s(x^s + 1.5y^s) + \lambda H^s x^s : \begin{array}{c} y^s \ge \omega^s - x^s \\ x^s, y^s \in \mathbb{Z}_+ \end{array}\right\}, \qquad \lambda \in \mathbb{R},$$

with $H^1 = 1$ and $H^2 = -1$.

- In the first iteration of the dual decomposition algorithm P_0 is the current problem. The optimal solution of the Lagrangian dual is $\lambda^\star = -0.25$ with optimal value $z_{LD} = 2.75$, and the first-stage part of the scenario solutions is $x^1 = 2$ and $x^2 = 0$. The (rounded) average solution is $\hat{x} = 1$, which has objective value $c\hat{x} + Q(\hat{x}) = 3.25 := \bar{z}$, the best value so far. Branching on x gives two new problems: P_1 with additional constraint $x \le 1$ and P_2 with additional constraint $x \ge 2$.
- In iteration 2 consider problem P_1. Since the solution $x = 1$ is already evaluated, the constraint $x \le 1$ can be replaced by $x = 0$. We find that $z_{LD} = 3.75 > \bar{z}$, so that P_1 is deleted from the list.
- The only remaining problem is P_2. For optimal $\lambda^\star \in [-0.5, -0.25]$ we find that $z_{LD} = 2.75 < \bar{z}$. The scenario solutions $x^1 = x^2 = 2$ agree, so that $\hat{x} = 2$ with objective value $c\hat{x} + Q(\hat{x}) = 2.75$ is an optimal first-stage solution. ◇

Structured Enumeration

An enumerative algorithm is due to Schultz et al. [42]. It applies to the MIR model with continuous first-stage and integer second-stage variables, (relatively) complete and sufficiently expensive recourse, and a finite discrete distribution of $h(\omega)$. The basic idea is to use structural properties of the expected value function Q to construct a countable (or even finite) subset V of the feasible set $X_0 := \{x \in X : Ax \le b\}$, that contains an optimal solution. In principle, for all x in this set the objective function $cx + Q(x)$ is then evaluated in order to identify an optimal solution.

First we briefly consider the computation of $Q(x)$ for all $x \in V$ as required by the algorithm. For every such x, this calls for the solution of all integer second-stage programming problems $\min\{qy : Wy \geq h^s - Tx,\ y \in \mathbb{Z}_+^p\}$, $s = 1, \ldots, S$. Since all these integer problems only differ in their right-hand side coefficients, it is suggested to solve them by means of a Gröbner basis, which is a particular *test set* (i.e. a finite set $y^1, \ldots, y^N$ of integral vectors, depending only on the matrix W, such that a feasible solution $y^\star$ is optimal if and only if $q(y^\star + y^k) \geq qy^\star$ whenever $y^\star + y^k$ is feasible, $k = 1, \ldots, N$, see e.g. [38]). This appears to be appropriate in this setting, since after computing the Gröbner basis which is independent of the right-hand side of the problem, solving each instance is very cheap. However, the algorithm does not depend on how the function evaluations are performed.

Since all second-stage variables are integral, all second-stage constraints need to be inequalities in order to obtain complete recourse as assumed (see Remark 4.1.4). Assuming that the recourse matrix W is integral (or rational, so that integrality can be obtained by scaling), it follows that for all non-negative integer vectors y, $Wy \geq t$ implies $Wy \geq \lceil t \rceil$, $t \in \mathbb{R}^m$. Therefore, the second-stage value function v is constant on subsets

$$\{t \in \mathbb{R}^m : \lceil t \rceil = k\} = \{t : k - (1, \ldots, 1)' < t \leq k\} \quad \forall k \in \mathbb{Z}^m,$$

which implies that for every $\bar{x} \in \mathbb{R}^n$ the expected value function Q is constant on

$$C(\bar{x}) = \bigcap_{s=1}^{S} \{x \in \mathbb{R}^n : \lceil h^s - Tx \rceil = \lceil h^s - T\bar{x} \rceil\}. \tag{4.15}$$

It is easy to see that the sets $C(\cdot)$ partition $\mathbb{R}^n$.

It follows that the objective function $cx + Q(x)$ is linear on each set $C(\cdot)$. Using that Q is lower semicontinuous (Theorem 4.2.3), it can be shown that the so-called set of candidates V, which consists of all vertices of the sets (4.15) intersected with X_0, contains an optimal solution.

Example 4.3.15 Consider the binary recourse problem

$$\max_x \{1.5x_1 + 4x_2 + Q(x) : x_j \in [-5, 5],\ j = 1, 2\},$$

where $Q(x) = \mathbb{E}_\omega [v(\omega - Tx)]$, and for $s \in \mathbb{R}^2$,

$$\begin{aligned} v(s) = \max_y\ & 16y_1 + 19y_2 + 23y_3 + 28y_4 \\ \text{s.t.}\ & 2y_1 + 3y_2 + 4y_3 + 5y_4 \leq s_1 \\ & 6y_1 + \ \ y_2 + 3y_3 + 2y_4 \leq s_2 \\ & y_j \in \{0, 1\},\ j = 1, \ldots, 4. \end{aligned}$$

The random variables $\omega_1 \in \{5.1\ 8.3\ 10.5\ 12.7\ 14.9\}$ and $\omega_2 \in \{5.0\ 8.9\ 10.6\ 12.4\ 14.3\}$ are independent with respective probabilities $\{0.0856\ 0.0184\ 0.2653\ 0.2654\ 0.3653\}$

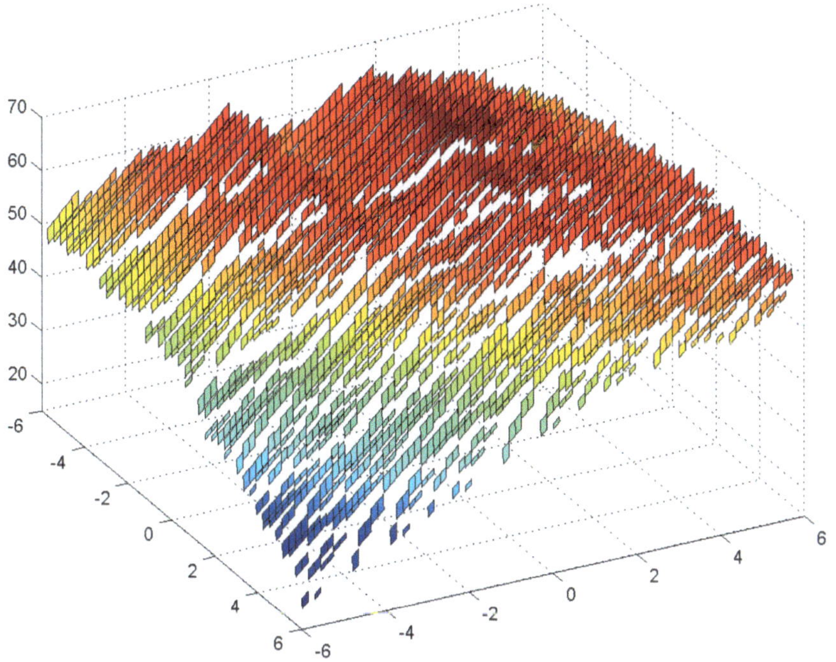

Fig. 4.13 The objective function $cx + Q(x)$ of Example 4.3.15

and {0.2105 0.2851 0.4561 0.0190 0.0293}. The deterministic matrix T equals $\begin{pmatrix} 1/3 & 2/3 \\ 2/3 & 1/3 \end{pmatrix}$. Figure 4.13 shows the piecewise linear, discontinuous objective function of this stochastic multi-knapsack problem. (Since this is a maximization problem, the objective function is *upper* semi-continuous.) ◇

Exercise 4.3.16 Verify that this stochastic multi-knapsack problem has relatively complete recourse. ◁

The set V is countable by construction, but it is finite only if X_0 is bounded. Since the goal is to find an optimal solution by enumeration, it would be useful if a finite subset of V can be determined that still contains an optimal solution. The following result can be used to this end.

Lemma 4.3.17 *Let D be a non-empty set, and f and $\bar{f}$ real functions on D such that $\bar{f}(x) \leq f(x)$ for all $x \in D$. Then, for all $\bar{x} \in D$,*

$$\operatorname*{argmin}_{x \in D} f(x) \subset \{x \in D : \bar{f}(x) \leq f(\bar{x})\}.$$

Moreover, the difference between these sets is smaller according as $\bar{f}$ is a better approximation of f and $f(\bar{x})$ is a better approximation of $\inf_{x\in D} f(x)$. *In particular, if $\bar{f}(\bar{x}) = f(\bar{x})$ and $\bar{x} \in \operatorname{argmin}_{x\in D} \bar{f}(x)$ then $\bar{x} \in \operatorname{argmin}_{x\in D} f(x)$.* □

Exercise 4.3.18 Prove Lemma 4.3.17. ◁

Now let $D = X_0$, $f(x) = cx + Q(x)$ and $\bar{f}$ the objective function of the continuous relaxation of MIR, that is, the continuous recourse problem that is obtained by dropping the integrality restrictions on the second-stage variables. Then $\bar{f} \le f$ (verify this), so that Lemma 4.3.17 implies that all optimal solutions of MIR are contained in the lower level set of $\bar{f}$ associated with an arbitrary feasible solution.

Exercise 4.3.19 Show that such level sets are bounded polyhedra if the recourse is complete and sufficiently expensive, and if the distribution of $h(\omega)$ is discrete. ◁

Thus, under the given assumptions on the recourse structure, the intersection of V with any such level set is a finite set that contains an optimal solution. Of course, this approach needs an explicit description of the level sets. Several algorithms for continuous recourse problems yield this information in the course of computation (see Sect. 3.4). For this reason, the continuous relaxation of MIR will be solved as the first step of the structured enumeration algorithm.

Since the cardinality of V is in general very high, it is important to identify subsets of candidate points that cannot be optimal solutions. To this end, two improvements of the algorithm are proposed. The first one consists of intersecting V with increasingly smaller level sets as suggested by Lemma 4.3.17. Secondly, based on the identification of common directions of increase of the first-stage objective cx and the expected value function $Q(x)$, certain groups of candidate points need not be evaluated. Numerical results for a small test problem are given.

Example 4.3.20 Consider the integer recourse problem

$$\min\{x + Q(x) : x \in \mathbb{R}_+ \},$$

where

$$\begin{aligned} Q(x) &= \mathbb{E}_\omega\left[v(x,\omega)\right] \\ v(x,\omega) &= \min\{1.5y : y \ge \omega - x,\ y \in \mathbb{Z}_+\} = 1.5\lceil \omega - x\rceil^+, \end{aligned}$$

and $\Pr\{\omega = 1.3\} = \Pr\{\omega = 2.7\} = 0.5$.

It is easy to see that the function $v(\cdot, \omega^s)$ is constant on every interval $[\omega^s + k, \omega^s + k + 1)$, $k \in \mathbb{Z}$, so that Q is constant on $[1.3+k, 1.7+k)$ and $[2.7+k, 3.3+k)$. Hence, the set of candidate points is

$$V = \left\{\{1.3, 2.7\} + \mathbb{Z}\right\} \cap \mathbb{R}_+ = \{0.3, 0.7, 1.3, 1.7, \ldots\}.$$

Solving the continuous relaxation of this problem (i.e. with second-stage value function $1.5(\omega - x)^+$), we learn that it has optimal solution $\bar{x} = 1.3$, and that its lower level sets are given by

$$L(\alpha) = \left\{ x \in \mathbb{R}_+ : \begin{array}{l} x \geq 6 - 2\alpha \\ x \leq 4\alpha - 8.1 \\ x \leq \alpha \end{array} \right\}$$

(verify this). The MIR objective value of $\bar{x}$ is 2.8, so that the associated level set is $[0.4, 2.8]$. By Lemma 4.3.17, the intersection of V with this level set, which is $\{0.7, 1.3, 1.7, 2.3, 2.7\}$, contains an optimal solution of MIR. Finally, using enumeration we find that $x^\star = 1.7$ is an optimal solution with value 2.46. ◇

Chapter 5
Chance Constraints

As has been indicated in Chap. 1, chance constraints arise as tools for modeling risk and risk aversion in random linear programs, interpreted as here-and-now decision problems. In this chapter we deal with the properties of these tools.

5.1 Modeling with Chance Constraints

As in Sect. 3.2, suppose that the following linear programming model with random parameters in the constraints has been formulated

$$(\mathrm{LP}_0(\omega)) \qquad \min_{x \in \mathbb{R}^n} \left\{ cx : \begin{array}{c} Ax = b \\ T(\omega)x \sim h(\omega) \\ x \in X \end{array} \right\}.$$

It is supposed that we are dealing with a decision problem, that is, one has to decide upon x before knowing the actual values of the random vector ω; only its distribution on Ω is supposed to be known. Consequently, $\mathrm{LP}_0(\omega)$ needs reformulation of the random constraints $T(\omega)x \sim h(\omega)$. One way to do this, is to require that these goal constraints are satisfied for sufficiently many realizations of ω. This is the rationale of chance constraints modeling. That is, $\mathrm{LP}_0(\omega)$ is interpreted as

$$(\mathrm{SLPwJCC}) \qquad \min_{x \in \mathbb{R}^n} \{cx : Ax = b,\ p(x) \geq \alpha,\ x \in X\}$$

with

$$p(x) := \Pr\{T(\omega)x \sim h(\omega)\}, \quad x \in \mathbb{R}^n,$$

and α some fixed number in $[0, 1]$, chosen by the decision maker.

W. K. Klein Haneveld et al., *Stochastic Programming*, Graduate Texts in Operations Research, https://doi.org/10.1007/978-3-030-29219-5_5

Here $p(x)$ is the *reliability* of the decision $x \in \mathbb{R}^n$, that is the probability that the goal constraint $T(\omega)x \sim h(\omega)$ is satisfied; its complement $1 - p(x)$ is the 'risk' of infeasibility associated with x. In model SLPwJCC the decision maker accepts possible violation of the goal constraints, but only if the risk is at most $1 - \alpha$. The number α is the minimum required reliability. Any x is called feasible (with respect to the goal constraint $T(\omega)x \sim h(\omega)$) if and only if its reliability is sufficiently large (i.e. at least α).

The choice of the value of α is at the discretion of the decision maker. The value $\alpha = 0$ corresponds to an extremely risky attitude, since $p(x) \geq 0$ for all $x \in \mathbb{R}^n$: the goal constraints are not taken care of at all. The value $\alpha = 1$ corresponds to an extremely conservative attitude. Such a value might look attractive from the point of view of risk aversion, but usually it is not; the reason is that more often than not there do not exist decisions x with $p(x) = 1$, so that SLPwJCC is an infeasible optimization problem, and it is of no help for finding suitable solutions for problems with uncertainty in the parameters. In practice, one usually takes α from $(0.5, 1.0)$. Statisticians may be inclined to take $\alpha = 0.95$ (or $\alpha = 0.90$, $\alpha = 0.975$). But this is not necessarily appropriate, since implicitly one is looking for a compromise with the objective to minimize cx, too (see, for example, the discussion after Exercise 1.2.12).

Remark 5.1.1 Robust optimization problems in which random constraints need to hold for all values of ω in a so-called uncertainty set can be interpreted as chance constraint problems with $\alpha = 1$. In robust optimization, however, the uncertainty set is typically carefully selected in order to avoid optimal solutions being too extremely conservative.

Up to now, we considered the m goal constraints $T(\omega)x \sim h(\omega)$ (recall that m is the number of rows in T and h) as one joint goal, leading to the *joint chance constraint*

$$p(x) := \Pr\{T(\omega)x \sim h(\omega)\} \geq \alpha.$$

It is also possible to deal with the goal constraints

$$T_i(\omega)x \sim h_i(\omega) \quad i = 1, \ldots, m,$$

separately. Then we define *individual reliabilities*

$$p_i(x) := \Pr\{T_i(\omega)x \sim h_i(\omega)\}, \ x \in \mathbb{R}^n, \quad i = 1, \ldots, m,$$

specify individual minimum required reliabilities $\alpha_i \in [0, 1]$, $i = 1, \ldots, m$, and agree that $x \in \mathbb{R}^n$ is called feasible (with respect to the goal constraint $T(\omega)x \sim h(\omega)$) if and only if

$$p_i(x) \geq \alpha_i \quad \text{for all } i = 1, \ldots, m.$$

Obviously, also intermediate specifications are possible, leading to chance constraints of the type

$$\Pr\{T_i(\omega)x \sim h_i(\omega),\ i \in I\} \geq \alpha_I$$

for some index set $I \subset \{1, \ldots, m\}$.

How to decide between joint or individual chance constraints? From the modeling point of view, if the individual goal constraints together describe one goal one will not be interested in the individual reliabilities of each constraint separately, and the joint chance constraint seems to be appropriate. On the other hand, if the individual goal constraints describe different goals, it seems to be appropriate to consider them separately. The use of individual chance constraints allows the decision maker to distinguish between crucial goals and less important goals, by selecting appropriate values for α_i, $i = 1, \ldots, m$. In that sense individual chance constraints are a more flexible modeling tool than joint chance constraints. In practice, other considerations are important, too. In order to make calculations, one has to come up with reduced forms of the chance constraints, and in general it is easier to derive reduced forms of individual chance constraints, or they are easier to deal with in optimization algorithms, than the reduced form of the corresponding joint chance constraint (see for instance the blending problem in Sect. 1.2.1). As a consequence, it is not unusual to deal with individual chance constraints, even if a joint chance constraints is more appropriate. That is, by replacing $p(x) \geq \alpha$ in SLPwJCC by $p_i(x) \geq \alpha_i$, $i = 1, \ldots, m$, for values of α_i to be chosen, one finds a suitable feasible solution that satisfies $p(x) \geq \alpha$.

Exercise 5.1.2 Show that $\alpha_i = 1 - \frac{1-\alpha}{m}$, $i = 1, \ldots, m$, is a suitable choice. (See Assignment C4.) ◁

We conclude this introduction with two remarks.

Remark 5.1.3 Suppose that one of the goal constraints $T(\omega)x \sim h(\omega)$ is an equality, actually:

$$T_i(\omega)x = h_i(\omega) \quad \text{for some } i \in \{1, \ldots, m\}.$$

Then, apart from uninteresting cases, we will have $p_i(x) = 0$ (hence $p(x) = 0$) if ω has a continuous distribution. Also for discrete distributions $p_i(x)$ and $p(x)$ will be close to zero, for all $x \in \mathbb{R}^n$. Consequently, *chance constraints are useless for equalities in the underlying goal constraints*, since they lead to infeasible problems for reasonable values of α (or α_i). In such cases one might consider relaxations

$$\Pr\{-a_i \leq T_i(\omega)x - h_i(\omega) \leq b_i\} \geq \alpha_i$$

for suitable positive a_i and b_i to be chosen. Usually, however, one restricts the use of chance constraints to one-sided goal constraints, say $T(\omega)x \geq h(\omega)$.

Remark 5.1.4 It might be appropriate to formulate *several chance constraints* for one individual (or joint) goal constraint. For instance, the goal $T_i(\omega)x \geq h_i(\omega)$ may be reformulated as

$$\Pr\{T_i(\omega)x \geq h_i(\omega) - d_{ik}\} \geq \alpha_{ik}, \quad k = 1, \ldots, K,$$

where $0 = d_{i1} < d_{i2} < \cdots < d_{iK}$ and $\alpha_{i1} < \alpha_{i2} < \cdots < \alpha_{iK}$. In this way one specifies that large shortages are disliked more than small ones. This approach is generalized in Chap. 6, where the *amount* of violation of the goal constraint is chosen as risk measure, leading to so-called *integrated* chance constraints (Sect. 6.1).

5.2 Mathematical Properties of Chance Constraints

Consider the model with a joint chance constraint as introduced in the previous section

$$\text{(SLPwJCC)} \qquad \min_{x \in \mathbb{R}^n} \{cx : Ax = b,\ p(x) \geq \alpha,\ x \in X\}.$$

Here, as before, X is a set of simple linear (in-)equalities, e.g. $X = \mathbb{R}^n_+$. The reliability function p is defined as

$$p(x) := \Pr\{T(\omega)x \geq h(\omega)\}, \quad x \in \mathbb{R}^n,$$

where, in accordance with Remark 5.1.3, we only start off with one-sided goal constraints. The number $\alpha \in [0, 1]$ is the minimum required (joint) reliability, chosen by the decision maker.

5.2.1 Different Representations of the Reliability Function

The reliability $p(x)$ can be represented as

$$p(x) = \Pr\{x \in S(\omega)\},$$

where $S(\omega)$ is the polyhedral convex set in $\mathbb{R}^n$ describing all decisions $x \in \mathbb{R}^n$ that satisfy the goal constraint if the realization of the random vector appears to be ω:

$$S(\omega) := \{x \in \mathbb{R}^n : T(\omega)x \geq h(\omega)\}.$$

This representation indicates that, in order to have $p(x) \geq \alpha$, the decision vector x should be an element of 'sufficiently many' sets $S(\omega)$. Alternatively, assuming as before (see Sect. 3.2) that $T(\omega)$ and $h(\omega)$ depend linearly on ω, we may write

$$p(x) = \Pr\{\omega \in K(x)\},$$

where $K(x)$ is a polyhedral convex set in $\mathbb{R}^r$, given by

$$K(x) := \left\{\omega \in \mathbb{R}^r : \sum_{k=1}^{r} \left(T^k x - h^k\right) \omega_k \geq -\left(T^0 x - h^0\right)\right\};$$

it describes all realizations of the random vector ω 'favorable' for x (i.e. $T(\omega)x \geq h(\omega)$).

We will use both representations of the reliability function p, depending on which representation is most convenient in each particular setting.

5.2.2 Examples of Chance Constraints

The optimization problem SLPwJCC is just a linear programming problem, apart from the constraint $p(x) \geq \alpha$. We are interested in properties of the corresponding feasible sets $C(\alpha)$, the upper level sets of p:

$$C(\alpha) := \left\{x \in \mathbb{R}^n : p(x) \geq \alpha\right\}, \quad \alpha \in [0, 1].$$

In case of individual chance constraints, the feasible set is

$$C(\alpha_1, \ldots, \alpha_m) := \bigcap_{i=1}^{m} C_i(\alpha_i), \quad C_i(\alpha_i) = \left\{x \in \mathbb{R}^n : p_i(x) \geq \alpha_i\right\}.$$

In particular, we are interested in the following questions.

(i) Are these feasible sets *closed*?
(ii) Are these feasible sets *convex*?

From optimization point of view, closed feasible sets are preferred. (Why?) Moreover, convexity is very convenient, since minimizing a linear function on a convex set is much easier than on a nonconvex set. Obviously, the answers to these questions depend on properties of the reliability functions $p(x)$, $p_i(x)$.

Before answering these questions, let us work out some simple examples. All these examples consider a single random constraint, and thus represent an individual chance constraint. We focus on whether the feasibility sets $C(\alpha)$ are closed, convex and/or non-empty.

Example 5.2.1 Consider the production planning problem from Sect. 1.2.2 in which $x \geq 0$ represents production, $\omega \sim F$ represents random demand, and the goal constraint is $x \geq \omega$. We require that demand is satisfied with at least probability α, leading to the chance constraint $\Pr\{x \geq \omega\} \geq \alpha$.

Simple computation yields

$$S(\omega) = [\omega, \infty) \quad \text{for all } \omega \in \mathbb{R},$$
$$K(x) = (-\infty, x] \quad \text{for all } x \in \mathbb{R},$$

so that using e.g. the second representation,

$$p(x) = \Pr\{\omega \in K(x)\} = \Pr\{\omega \leq x\} = F(x), \quad x \in \mathbb{R}.$$

It follows directly that the feasibility set $C(\alpha)$ is given by

$$C(\alpha) = \left[F^{-1}(\alpha), \infty\right), \quad \alpha \in [0, 1],$$

where

$$F^{-1}(\alpha) := \inf_{t \in [-\infty, \infty)} \{t : F(t) \geq \alpha\}.$$

In other words, a production level x is feasible, i.e. $x \in C(\alpha)$, if and only if x exceeds the α-quantile $F^{-1}(\alpha)$ of the demand distribution of ω.

We conclude that in this case, *for all* $\alpha \in [0, 1]$ and *all distributions* F, the feasible set $C(\alpha)$ is closed and convex. Moreover, it is nonempty, except maybe for $\alpha = 1$. ◇

It turns out, however, that $C(\alpha)$ is not necessarily convex for all $\alpha \in [0, 1]$, as we will illustrate in the next example.

Example 5.2.2 Consider the blending problem from Sect. 1.2.1 in which $x \in \mathbb{R}^2_+$ represent the mix of ingredients and $\omega \in \mathbb{R}^2$ the uncertain nutritious value of ingredient 1. We assume that the blending problem has been modeled using individual chance constraint, and consider only the first:

$$\mathbb{P}\{\omega_1 x_1 + x_2 \geq 7\} \geq \alpha_1.$$

For exposition purposes we write $\omega := \omega_1$ and $\alpha := \alpha_1$ in the remainder of this example, and we assume that x may attain negative values. The cdfs of ω are denoted by F and $\tilde{F}$ (i.e., $F(t) = \Pr\{\omega \leq t\}$ and $\tilde{F}(t) = \Pr\{\omega < t\}$, $t \in \mathbb{R}$).

Then,

$$S(\omega) = \left\{x \in \mathbb{R}^2 : \omega x_1 + x_2 \geq 7\right\},$$

$$K(x) = \begin{cases} \left[\frac{7-x_2}{x_1}, \infty\right), & x_1 > 0 \\ \emptyset, & x_1 = 0 \text{ and } x_2 < 7 \\ \mathbb{R}, & x_1 = 0 \text{ and } x_2 \geq 7 \\ \left(-\infty, \frac{7-x_2}{x_1}\right], & x_1 < 0. \end{cases}$$

Here, $K(x)$ is derived using $\omega x_1 + x_2 > 7 \Leftrightarrow \omega > \frac{7-x_2}{x_1}$ if $x_1 > 0$ and $\omega x_1 + x_2 \geq 7 \Leftrightarrow \omega \leq \frac{7-x_2}{x_1}$ if $x_1 < 0$. Using the expression for $K(x)$, we easily derive that

$$p(x) = \begin{cases} 1 - \tilde{F}\left(\frac{7-x_2}{x_1}\right), & x_1 > 0 \\ 0, & x_1 = 0 \text{ and } x_2 < 7 \\ 1, & x_1 = 0 \text{ and } x_2 \geq 7 \\ F\left(\frac{7-x_2}{x_1}\right), & x_1 < 0. \end{cases}$$

Partitioning the feasibility set $C(\alpha)$ into $C_+(\alpha)$, $C_0(\alpha)$, and $C_-(\alpha)$, corresponding to $x_1 < 0$, $x_1 = 0$, and $x_1 > 0$, respectively, we conclude that $C(\alpha) = C_+(\alpha) \cup C_0(\alpha) \cup C_-(\alpha)$, where, for $0 < \alpha < 1$

$$C_+(\alpha) := \left\{(x_1, x_2) \in \mathbb{R}^2 : x_1 > 0,\ \tilde{F}^{-1}(1-\alpha)x_1 + x_2 \geq 7\right\}$$

$$C_0(\alpha) := \left\{(0, x_2) \in \mathbb{R}^2 : x_2 \geq 7\right\}$$

$$C_-(\alpha) := \left\{(x_1, x_2) \in \mathbb{R}^2 : x_1 < 0,\ F^{-1}(\alpha)x_1 + x_2 \geq 7, \right\}$$

where $F^{-1}(\alpha)$ is defined in Example 5.2.1 and

$$\tilde{F}^{-1}(1-\alpha) := \max_{t \in (-\infty,\infty)} \{t : \tilde{F}(t) \leq 1 - \alpha\}.$$

We conclude that for *all values* of $\alpha \in (0, 1)$ and *for all distributions of* ω, the set $C(\alpha)$ is closed. Figure 5.1 shows, however, that it is only convex if $\alpha \geq 0.5$, for all distributions of ω. ◇

The next example is a special case of Example 5.2.2 with $x_2 = 0$. It shows that it is not trivial whether and when $C(\alpha)$ is non-empty.

Example 5.2.3 Let $T(\omega) = \omega$ and $h(\omega) = 1$, where ω is a random variable with cdfs F and $\hat{F}$ (i.e., $F(t) = \Pr\{\omega \leq t\}$ and $\tilde{F}(t) = \Pr\{\omega < t\}$, $t \in \mathbb{R}$). The goal

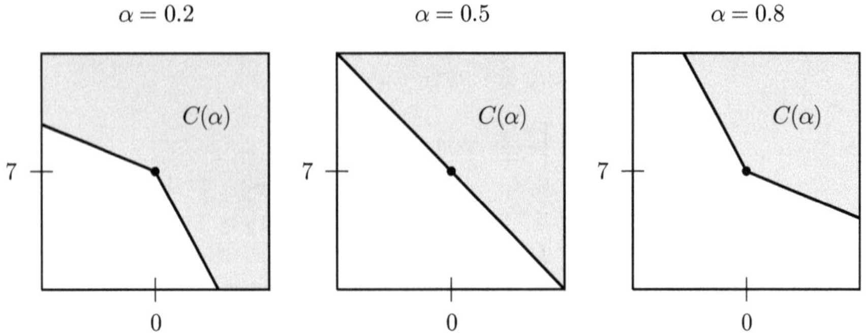

Fig. 5.1 The feasible set $C(\alpha)$ of Example 5.2.2 with ω uniformly distributed on [0, 1] for $\alpha =$ 0.2, 0.5, and 0.8

constraint $T(\omega)x \geq h(\omega)$ is now

$$\omega x \geq 1$$

so that

$$S(\omega) = \{x \in \mathbb{R} : \omega x \geq 1\}, \quad \omega \in \mathbb{R},$$

$$K(x) = \begin{cases} \left(-\infty, \frac{1}{x}\right], & x < 0 \\ \emptyset, & x = 0 \\ \left[\frac{1}{x}, \infty\right), & x > 0, \end{cases}$$

and

$$p(x) = \begin{cases} F\left(\frac{1}{x}\right), & x < 0 \\ 0, & x = 0 \\ 1 - \tilde{F}\left(\frac{1}{x}\right), & x > 0. \end{cases}$$

Hence, if ω is a nonnegative random variable (i.e. $\tilde{F}(0) = 0$), then

$$C(\alpha) = \left[\left(\tilde{F}^{-1}(1-\alpha)\right)^{-1}, \infty\right)$$

and if ω is nonpositive (i.e. $F(0) = 1$), then

$$C(\alpha) = \left(-\infty, \left(F^{-1}(\alpha)\right)^{-1}\right].$$

In both cases $C(\alpha)$ is a closed convex set for all $\alpha \in [0, 1]$, but empty for $\alpha \geq 1 - \Pr\{\omega = 0\}$. On the other hand, if $\Pr\{\omega > 0\} > 0$ and $\Pr\{\omega < 0\} > 0$, then

$$\sup_{x<0} p(x) = \tilde{F}(0) \quad \text{and} \quad \sup_{x>0} p(x) = 1 - F(0).$$

Hence, if $\alpha > \max\{\tilde{F}(0), 1 - F(0)\}$, then $C(\alpha)$ is empty and if $0 < \alpha < \min\{\tilde{F}(0), 1 - F(0)\}$, then

$$C(\alpha) = \left(-\infty, \left(F^{-1}(\alpha)\right)^{-1}\right] \cup \left[\left(\tilde{F}^{-1}(1-\alpha)\right)^{-1}, \infty\right)$$

is closed but nonconvex. In particular, if ω is a continuous random variable with bounded support whose distribution is symmetrical around 0, we get

$$C(\alpha) = \begin{cases} \mathbb{R}, & \alpha = 0 \\ (-\infty, -A] \cup [A, \infty),\ A := -\left(F^{-1}(\alpha)\right)^{-1} > 0, & 0 < \alpha < 1/2 \\ \emptyset, & \alpha \in [1/2, 1], \end{cases}$$

so that the set $C(\alpha)$ is nonconvex for small values of α, whereas it is empty for all relevant values of α! ◇

Remark 5.2.4 For more examples of individual CC's with one random variable, see Assignment C6.

In the previous examples ω was a one-dimensional random variable with arbitrary distribution function. In the next examples, however, ω will be a random vector (of dimension 2) following a discrete distribution. We will see that in this case it will be more convenient to determine $p(x)$ using $S(\omega)$ than $K(x)$. Moreover, also in these examples we see that $C(\alpha)$ may be non-convex.

Example 5.2.5 Let $(T(\omega), h(\omega)) = (\omega_1, \omega_2)$ where $\omega = (\omega_1, \omega_2)$ is discretely distributed: $\Pr\{(\omega_1, \omega_2) = (-2, -1)\} = 1/3$, $\Pr\{(\omega_1, \omega_2) = (2, 2)\} = 2/3$. The goal constraint $T(\omega)x \geq h(\omega)$ is now

$$\omega_1 x \geq \omega_2$$

so that

$$S(\omega) = \begin{cases} (-\infty, 1/2], & \omega = (-2, -1) \\ [1, \infty), & \omega = (2, 2), \end{cases}$$

$$K(x) = \left\{\omega \in \mathbb{R}^2 : x\omega_1 - \omega_2 \geq 0\right\}, \quad x \in \mathbb{R}.$$

Using $p(x) = \Pr\{x \in S(\omega)\}$, we conclude that

$$p(x) = \begin{cases} 1/3, & x \in (-\infty, 1/2] \\ 0, & x \in (1/2, 1) \\ 2/3, & x \in [1, \infty), \end{cases}$$

$$C(\alpha) = \begin{cases} \mathbb{R}, & \alpha = 0 \\ (-\infty, 1/2] \cup [1, \infty), & 0 < \alpha \leq 1/3 \\ [1, \infty), & 1/3 < \alpha \leq 2/3 \\ \emptyset, & 2/3 < \alpha \leq 1. \end{cases}$$

Here $C(\alpha)$ is convex for all $\alpha > 1/3$, but empty for $\alpha > 2/3$. For all values of α it is a closed set, though. ◇

Example 5.2.6 Let $T(\omega) = (-\omega_1, -\omega_2)$ and $h(\omega) = -1$, where $\omega = (\omega_1, \omega_2)$ follows the discrete distribution $\Pr\{\omega = (3, 0)\} = 1/7$, $\Pr\{\omega = (0, 3)\} = 2/7$, $\Pr\{\omega = (1, 1)\} = 4/7$. The goal constraint $T(\omega)x \geq h(\omega)$ is now

$$\omega_1 x_1 + \omega_2 x_2 \leq 1,$$

so that

$$S(\omega) = \begin{cases} \left\{x \in \mathbb{R}^2 : x_1 \leq 1/3\right\}, & \omega = (3, 0) \\ \left\{x \in \mathbb{R}^2 : x_2 \leq 1/3\right\}, & \omega = (0, 3) \\ \left\{x \in \mathbb{R}^2 : x_1 + x_2 \leq 1\right\}, & \omega = (1, 1), \end{cases}$$

$$K(x) = \left\{\omega \in \mathbb{R}^2 : x_1\omega_1 + x_2\omega_2 \leq 1\right\}, \quad x \in \mathbb{R}^2.$$

The various levels of $p(x)$ are denoted in Fig. 5.2. At points where p is discontinuous, its value is equal to the maximum value in the neighborhood, as can be verified directly. From the figure it is easy to characterize the feasible set $C(\alpha)$ for all $\alpha \in [0, 1]$. All of them are nonempty and closed, but convexity is only present for $\alpha = 0$ (trivially), $2/7 < \alpha \leq 4/7$, and $5/7 < \alpha \leq 1$. ◇

The specification of $C(\alpha)$ with discretely distributed ω, as in the last two examples, is formulated in the next exercise. More details are given in Sect. 5.3.

Exercise 5.2.7 Verify that, at least if ω follows a finite discrete distribution,

$$C(\alpha) = \bigcup_{K \in \mathcal{K}_\alpha} \bigcap_{\omega \in K} S(\omega), \quad \alpha \in [0, 1],$$

where $\mathcal{K}_\alpha := \{K \subset \Omega : \Pr\{\omega \in K\} \geq \alpha\}$ and $S(\omega) := \{x \in \mathbb{R}^n : T(\omega)x \geq h(\omega)\}$. Observe that $C(\alpha)$ is a union of convex sets, which is in general not convex. This representation of $C(\alpha)$ is the basis of the analysis in Sect. 5.3. ◁

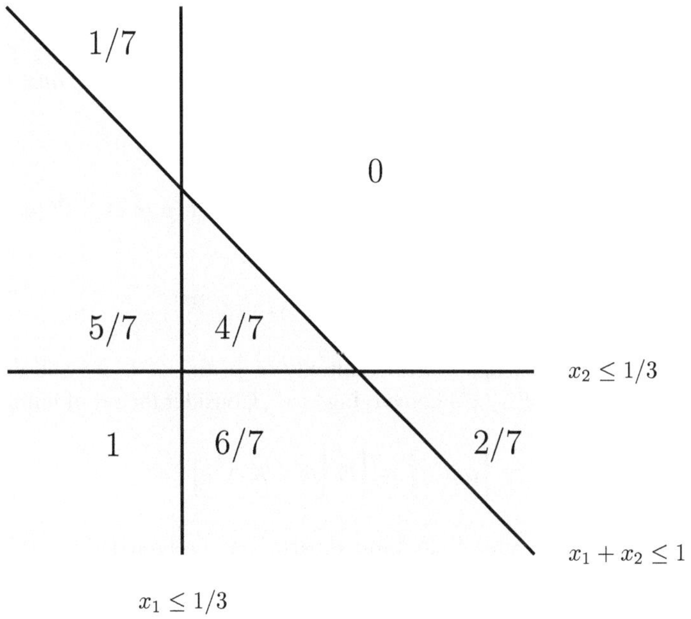

Fig. 5.2 Levels of the reliability function p of Example 5.2.6. The shaded region corresponds to the feasibility set $C(\alpha)$ for $1/7 < \alpha \leq 2/7$

The examples in this section suggest that the feasible sets $C(\alpha)$ are closed, possibly empty, but not necessarily convex. This is supported by a formal analysis in the next section.

5.2.3 *Properties of Chance Constraints*

Closedness and Monotonicity of Feasibility Sets

Since $C(\alpha)$, $\alpha \in [0, 1]$ are all upper level sets of the reliability function p, all of them are closed if and only if p is an *upper semicontinuous* (usc) function, that is:

$$\forall x \in \mathbb{R}^n : \limsup_{y \to x} p(y) \leq p(x).$$

It appears, that without conditions any reliability function p is usc.

Theorem 5.2.8

(a) For all reliability functions p it holds that

$$p(x) \geq \limsup_{y \to x} p(y), \quad x \in \mathbb{R}^n. \tag{5.1}$$

That is, p is usc and hence $C(\alpha)$ is a closed set for all $\alpha \in [0, 1]$.

(b) *As a multifunction of $\alpha \in [0, 1]$, $C(\alpha)$ is nonincreasing: if $0 \le \alpha_1 < \alpha_2 \le 1$ then $C(\alpha_1) \supset C(\alpha_2)$. $C(0) = \mathbb{R}^n$, $C(\alpha) \neq \emptyset$ for all $\alpha \in [0, 1]$ if and only if the 'extremely safe set' $C(1)$ is not empty.*

Proof

(a) Define $f(x, \omega) = T(\omega)x - h(\omega)$, $x \in \mathbb{R}^n$, $\omega \in \Omega$. Then $p(x) = \Pr\{\omega \in K(x)\}$, where

$$K(x) = \left\{\omega \in \Omega : f(x, \omega) \in \mathbb{R}^m_+\right\}.$$

In order to establish upper semicontinuity of p, fix $x^* \in \mathbb{R}^n$ and define any sequence $\{x^k\} \subset \mathbb{R}^n \setminus \{x^*\}$ converging to x^*. Consider the set of numbers

$$\left\{p(x^k)\right\} = \left\{\Pr\left\{\omega \in K(x^k)\right\}\right\}.$$

According to Appendix A, its limit superior can be bounded by $\Pr\left\{\omega \in \overline{K}\right\}$, where

$$\begin{aligned}\overline{K} &= \left\{\omega \in \Omega : \omega \in K(x^k) \text{ for infinitely many } k\right\} \\ &= \left\{\omega \in \Omega : f(x^k, \omega) \in \mathbb{R}^m_+ \text{ for infinitely many } k\right\}.\end{aligned}$$

We will show that $\overline{K} \subset K(x^*)$. Fix $\omega \in \overline{K}$. Obviously, there is a subsequence $\{x^{k_j}\}$ such that $f(x^{k_j}, \omega) \in \mathbb{R}^m_+$ for all $j = 1, 2, \ldots$ Since $x^{k_j} \to x^*$ and f is continuous in x for every ω, it follows that $f(x^{k_j}, \omega) \to f(x^*, \omega)$ as $x^{k_j} \to x^*$. Moreover, since $\mathbb{R}^m_+$ is closed, it follows that $f(x^*, \omega) \in \mathbb{R}^m_+$, so that $\omega \in K(x^*)$ indeed. This completes the proof of $\overline{K} \subset K(x^*)$.

Consequently,

$$\limsup_{k \to \infty} p(x^k) \le \Pr\left\{\omega \in \overline{K}\right\} \le \Pr\left\{\omega \in K(x^*)\right\} = p(x^*).$$

Since the inequality is true for all such sequences $\{x^k\}$, we conclude that

$$\limsup_{y \to x^*} p(y) \le p(x^*).$$

(b) Obvious.

□

Remark 5.2.9 In many cases (5.1) actually holds with equality. However, there are examples such that the inequality is strict. Consider, for example, the reliability function p of Example 6.2.2 for $x = 0$.

Convex Feasibility Sets for Fixed T

Next, we consider the convexity of $C(\alpha)$. Since $C(\alpha)$, $\alpha \in [0, 1]$, are all upper level sets of p, all of them are convex if and only if p is a *quasi-concave* function (defined in Appendix B), i.e. if $x^0 \in \mathbb{R}^n$, $x^1 \in \mathbb{R}^n$, $\lambda \in [0, 1]$, and $x^\lambda := (1-\lambda)x^0 + \lambda x^1$ imply $p(x^\lambda) \geq \min\{p(x^0), p(x^1)\}$. The examples in Sect. 5.2.2 show that $C(\alpha)$ is not necessarily convex, and thus the reliability function is not necessarily quasi-concave.

In special cases, however, the reliability function is quasi-concave indeed. For instance, consider the case that $T(\omega) = T$ is fixed, that is, the goal constraint is

$$Tx \geq \omega$$

where ω is an m-dimensional random vector with cdf F. Now $p(x) = F(Tx)$, so that p is quasi-concave if F is (only if holds too if rank $T = m$). Therefore, we consider the question: which distribution functions are quasi-concave. Obviously, any concave function is quasi-concave, but that does not help much since there does not exist a cdf F that is concave on $\mathbb{R}^n$. (Why?) Fortunately, there are many quasi-concave functions that are not concave. For instance, every *univariate* cdf F is quasi-concave, because it is nondecreasing, and every multivariate cdf F is quasi-concave if F is *log-concave* (i.e., if $\log F$, with $\log 0 := -\infty$, is an extended real concave function on $\mathbb{R}^n$). The first observation implies that individual chance constraints with fixed T must give rise to convex feasibility sets.

Theorem 5.2.10 *Let T be a constant $1 \times n$ row vector and ω a random variable with cdf F, and consider the random goal constraint $Tx \geq \omega$. Then, $C(\alpha)$ is a closed convex set for all $\alpha \in [0, 1]$. In fact, it has the following linear reduced form:*

$$C(\alpha) = \left\{x \in \mathbb{R}^n : Tx \geq F^{-1}(\alpha)\right\}, \quad \alpha \in (0, 1),$$

where $F^{-1}(\alpha) := \min_{t \in \mathbb{R}}\{t : F(t) \geq \alpha\}$, $\alpha \in (0, 1)$. □

The second observation implies that joint chance constraints with fixed T may also yield convex feasibility sets, at least, if the cdf F of ω is log-concave.

Theorem 5.2.11 *Let T be a constant $m \times n$ matrix and ω an m-dimensional random vector with cdf F, and consider the random goal constraint $Tx \geq \omega$. If F is log-concave, then $C(\alpha)$ is a closed convex set for all $\alpha \in [0, 1]$.* □

Exercise 5.2.12 Prove Theorem 5.2.11. ◁

Exercise 5.2.13 More general, assume that a strictly increasing extended real function $\varphi : [0, 1] \mapsto [-\infty, \infty]$ exists such that the composite function $\varphi(F(x))$ is concave. Prove that F is quasi-concave. ◁

The following theorem, combining results of Prékopa and Borell (see [33]), considers continuous random vectors ω and provides sufficient conditions on multivariate *density functions* f such that the feasibility set $C(\alpha)$ is convex for all $\alpha \in [0, 1]$. The theorem is given without proof.

Theorem 5.2.14 *Suppose that T is a constant $m \times n$ matrix, $m \geq 2$, and ω is a continuously distributed m-dimensional random vector with pdf f and cdf F.*

(i) (Prékopa) If $\log f$ is concave (with $\log 0 = -\infty$), or
(ii) (Borell) If $f^{-1/m}$ is convex (with $0^{-1/m} = \infty$),

then F is quasi-concave, and hence $C(\alpha)$ is closed convex for all $\alpha \in [0, 1]$. □

The conditions (i) or (ii) in Theorem 5.2.14 are satisfied by many densities. For example, the uniform distribution on $\Omega \subset \mathbb{R}^m$ with Ω convex, the multivariate normal distribution, and the multivariate Student-distribution all have quasi-concave distribution functions.

Exercise 5.2.15 Verify this. ◁

Remark 5.2.16 It is not difficult to see that a density can be log-concave only if it is unimodal.

Convex Feasibility Sets for Random T

Next we consider a few special cases in which (also) elements of the technology matrix $T(\omega)$ are random. They all apply to individual chance constraints, hence below T is a $1 \times n$ vector. In the first special case, the right-hand side h is fixed and the vector T contains only one random element, and in the other special cases the random vector ω is *normally distributed*.

The first result was already illustrated in Example 5.2.2.

Theorem 5.2.17 *Suppose $h(\omega) = h$ is fixed and $T(\omega) = (\omega, t_2, \dots, t_n)$, where the random variable ω follows an arbitrary distribution. Then $C(\alpha)$ is convex if and only if $\alpha = 0$ or $\alpha \in [1/2, 1]$.* □

Exercise 5.2.18 Prove Theorem 5.2.17. (See Assignment C6.) ◁

Next, assume that ω is a random vector following a normal distribution. The following result was first proved by Katakao [15] and Van de Panne and Popp [46].

Theorem 5.2.19 *Suppose $h(\omega) = h$ is fixed and $T(\omega) = (\omega_1, \dots, \omega_n)$, where $\omega = (\omega_1, \dots, \omega_n)$ follows a normal distribution with mean $\mu = (\mu_1, \dots, \mu_n)$ and covariance matrix V. Then*

$$C(\alpha) = \left\{x \in \mathbb{R}^n : \mu' x \geq h + \Phi^{-1}(\alpha)(x' V x)^{1/2}\right\}, \quad \alpha \in [0, 1],$$

where $\Phi^{-1}(\alpha)$ is the α-quantile of the standard normal distribution. Moreover, the set $C(\alpha)$ is convex if $\alpha \geq 1/2$.

Proof The random variable $\omega' x$ follows a normal distribution with mean $\mu' x$ and variance $\sigma^2(x) = x' V x$. Hence

$$\begin{aligned}\Pr\left\{\omega' x \geq h\right\} &= \Pr\left\{\frac{\omega' x - \mu' x}{\sigma(x)} \geq \frac{h - \mu' x}{\sigma(x)}\right\} \\ &= 1 - \Phi\left(\frac{h - \mu' x}{\sigma(x)}\right) = \Phi\left(\frac{\mu' x - h}{\sigma(x)}\right),\end{aligned}$$

so that

$$\begin{aligned}C(\alpha) &= \left\{x \in \mathbb{R}^n : \Phi\left(\frac{\mu' x - h}{\sigma(x)}\right) \geq \alpha\right\} \\ &= \left\{x \in \mathbb{R}^n : \mu' x \geq h + \Phi^{-1}(\alpha)\sigma(x)\right\}.\end{aligned}$$

It remains to show that $C(\alpha)$ is convex if $\alpha \geq 1/2$. This follows from the observation that $\Phi^{-1}(\alpha) \geq 0$ in this case, and that $\sigma(x) = (x' V x)^{1/2}$ is a convex function of x, as we will show now. Since the covariance matrix V is positive semi-definite, there exists a square matrix C such that $C' C = V$. Therefore,

$$(x' V x)^{1/2} = (x' C' C x)^{1/2} = \|Cx\|_2,$$

which is a convex function indeed. □

Exercise 5.2.20 Implicitly, it is assumed that $\sigma(x) > 0$. Under which circumstances is $\sigma(x) = 0$? Show that the theorem is true under all circumstances. ◁

Corollary 5.2.21 *Suppose that $T(\omega) = (\omega_1, \ldots, \omega_n)$ and $h(\omega) = \omega_{n+1}$, where $\omega = (\omega_1, \ldots, \omega_{n+1})$ follows a normal distribution with mean $\mu = (\mu_1, \ldots, \mu_{n+1})$ and covariance matrix V. Then the set $C(\alpha)$ is convex if $\alpha \geq 1/2$.*

Proof By Theorem 5.2.19 the $(n+1)$-dimensional set $\bar{C}(\alpha)$,

$$\bar{C}(\alpha) = \left\{x \in \mathbb{R}^{n+1} : \Phi\left(\frac{\mu' x}{\sigma(x)}\right) \geq \alpha\right\}$$

is convex if $\alpha \geq 1/2$. The result follows, since $C(\alpha)$ is the intersection of $\bar{C}(\alpha)$ with the hyperplane $\{x \in \mathbb{R}^{n+1} : x_{n+1} = -1\}$, which is convex by Theorem B.11. □

Remark 5.2.22 The case $n = 1$ is dealt with in detail in Assignment C3.

Exercise 5.2.23 Formulate the reduced form of the set $C(\alpha)$ in Corollary 5.2.21, and give an interpretation of the result. ◁

5.3 Discrete Distributions

In this section we consider chance constrained problems for the case that ω follows a finite discrete distribution, say $\Pr\left\{\omega = \omega^k\right\} = p_k,\ k = 1, \ldots, K,\ \sum_k p_k = 1$. Even if ω is continuously distributed, from a computational point of view it may be necessary to approximate its support by a finite set of sample values in order to get a tractable problem.

Defining $S_k := S(\omega^k)$, where $S(\omega) = \{x \in \mathbb{R}^n : T(\omega)x \geq h(\omega)\}$ as before,

$$p(x) = \Pr\{x \in S(\omega)\} = \sum_{k=1}^{K} p_k \cdot 1_{S_k}(x).$$

From this representation we see that $p(x) \geq \alpha$ if x is in 'sufficiently many' sets S_k, meaning that $x \in S_k$ for all k in some $I \subset \{1, \ldots, K\}$ with $\sum_{k \in I} p_k \geq \alpha$. Since x is feasible if it belongs to any such set I, we obtain

$$C(\alpha) = \bigcup_{I \in \mathcal{N}_\alpha} \bigcap_{k \in I} S_k, \qquad \mathcal{N}_\alpha := \left\{ I \subset \{1, \ldots, K\} : \sum_{k \in I} p_k \geq \alpha \right\}.$$

Note that each set $\bigcap_{k \in I} S_k$ is convex polyhedral, but that the set $C(\alpha)$ is in general not convex since it is a union of (convex) sets. However, for a fixed α, the set $C(\alpha)$ is convex if there exists a *dominating index set* $I_\alpha \in \mathcal{N}_\alpha$, satisfying $I_\alpha \subset I$ for all $I \in \mathcal{N}_\alpha$. Such a set I_α is *minimal with respect to inclusion*.

Exercise 5.3.1 Verify that, for all $I \in \mathcal{N}_\alpha$,

$$\bigcap_{k \in I} S_k \subset \bigcap_{k \in I_\alpha} S_k$$

if I_α is a dominating index set. ◇

We conclude that $C(\alpha) = \bigcap_{k \in I_\alpha} S_k$ is convex polyhedral in that case.

Example 5.3.2 As in Example 5.2.6, consider the goal constraint $\omega_1 x_1 + \omega_2 x_2 \leq 1$, that is, $(T(\omega), h(\omega)) = (-\omega_1, -\omega_2, -1)$, and let $\omega = (\omega_1, \omega_2)$ be discretely distributed on three points: $\Pr\{\omega = (3, 0)\} = 1/7 = p_1$, $\Pr\{\omega = (0, 3)\} = 2/7 = p_2$, $\Pr\{\omega = (1, 1)\} = 4/7 = p_3$. The following table contains all subsets of the index set $\{1, 2, 3\}$ and the associated total probability, in increasing order from left to right.

I	$\emptyset$	$\{1\}$	$\{2\}$	$\{1, 2\}$	$\{3\}$	$\{1, 3\}$	$\{2, 3\}$	$\{1, 2, 3\}$
$\sum_{k \in I} p_k$	0	1/7	2/7	3/7	4/7	5/7	6/7	1

Using this list it is easy to see that $\emptyset$ and $\{3\}$ are dominating index sets for $\alpha = 0$ and $\alpha \in (3/7, 4/7]$, respectively. For example, for $\alpha = 4/7$, $\mathcal{N}_\alpha$ consists of the index sets $\{3\}$, $\{1, 3\}$, $\{2, 3\}$ and $\{1, 2, 3\}$, so that

$$C(4/7) = S_3 \cup (S_1 \cap S_3) \cup (S_2 \cap S_3) \cup (S_1 \cap S_2 \cap S_3) = S_3.$$

From Example 5.2.6 we know that $C(\alpha)$ is convex for $\alpha = 0$, $2/7 < \alpha \leq 4/7$, and $5/7 < \alpha \leq 1$, illustrating that existence of a dominating index set for some α is sufficient but not necessary for convexity of $C(\alpha)$. ◇

Remark 5.3.3 We refer to Assignment C5 for a treatment of the special case 'T is fixed'.

Thus, the set $C(\alpha)$ is in general non-convex if ω is discretely distributed. In this case, however, it is possible to formulate the chance constrained problem as a mixed-integer linear program. Indeed, if the distribution is given by $\Pr\left\{(T(\omega), h(\omega)) = (T^k, h^k)\right\} = p^k, k = 1, \ldots, K$, then

$$\min_{x \in \mathbb{R}^n} \{cx : \Pr\{T(\omega)x \geq h(\omega)\} \geq \alpha,\ x \in X\}$$

is equivalent to

$$\begin{aligned} \min_{x \in \mathbb{R}^n} \ & cx \\ \text{s.t.} \ & T_i^k x + M_i \delta^k \geq h_i^k, \quad k = 1, \ldots, K,\ i = 1, \ldots, m \\ & \sum_{k=1}^{K} p^k \delta^k \leq 1 - \alpha, \\ & x \in X, \quad \delta^k \in \{0, 1\}, \quad k = 1, \ldots, K \end{aligned} \tag{5.2}$$

Here M_i is a sufficiently large number, and the indicator δ^k is equal to zero if and only if $T^k x \geq h^k$. (The 'if' part holds under the assumption that any solution with $T^k x \geq h^k$ and $\delta^k = 1$ is not optimal.)

As can be expected from the structure of (5.2), it is possible to deal with chance constraints with discrete random variables in the framework of recourse models, if one allows for mixed-integer recourse. To illustrate this, consider the MIR model of Sect. 4.1 with all random variables discretely distributed. If one takes $p = m + 1$, $\bar{p} = 1$, $q = (q_0, 0)$ with $q_0 \in \mathbb{R}_+$, $W = (M, -I)$ where $M \in \mathbb{R}^m$ has sufficiently large elements M_i and I is the $m \times m$ identity matrix, and if the recourse variables are $y = (\delta, \bar{y})$ with $\delta \in \{0, 1\}$ and $\bar{y} \in \mathbb{R}^m_+$, then the MIR model (in canonical form) is

$$\min_{x \in \mathbb{R}^n} \left\{ cx + \sum_{k=1}^{K} p^k v(x, \omega^k) : x \in X \right\}$$

with

$$v(x, \omega^k) = \min \left\{ q_0 \cdot \delta : \begin{array}{c} M\delta - \bar{y} + T^k x = h^k \\ \delta \in \{0, 1\}, \quad \bar{y} \in \mathbb{R}^m_+ \end{array} \right\},$$

which is equivalent to

$$\min_{x \in \mathbb{R}^n} \{cx + q_0 \cdot \Pr\{T(\omega)x \not\geq h(\omega)\} : x \in X\}.$$

Example 5.3.4 Consider once more the goal constraint $\omega_1 x_1 + \omega_2 x_2 \leq 1$, with ω discretely distributed: $\Pr\{\omega = (3, 0)\} = 1/7$, $\Pr\{\omega = (0, 3)\} = 2/7$, $\Pr\{\omega = (1, 1)\} = 4/7$. Suppose that the aim is to minimize a weighted sum of costs cx and the probability that the goal constraint is not met, i.e.,

$$\min \left\{ cx + q_0 \cdot \Pr\{\omega_1 x_1 + \omega_2 x_2 > 1\} : x \in \mathbb{R}^2_+ \right\}.$$

This is equivalent to the MIR model

$$\min \left\{ cx + \mathbb{E}_\omega [v(x, \omega)] : x \in \mathbb{R}^2_+ \right\}$$

with

$$v(x, \omega) = \min \left\{ q_0 \cdot \delta : -M\delta + \omega_1 x_1 + \omega_2 x_2 \leq 1, \ \delta \in \{0, 1\} \right\},$$

where M is a sufficiently large number. ◇

5.4 Algorithms for Models with a Joint Chance Constraint

From a theoretical point of view, a model with a joint chance constraint SLPwJCC is a special case of a non-linear convex optimization problem, at least if the feasible set $C(\alpha) = \{x \in \mathbb{R}^n : \Pr\{T(\omega)x \geq h(\omega)\} \geq \alpha\}$ is convex. Indeed, in that case the problem SLPwJCC is

$$\min_{x \in \mathbb{R}^n} \{cx : Ax \sim b, \ G(x) \leq 0, \ x \in X\},$$

where $G(x) := \alpha - \Pr\{T(\omega)x \geq h(\omega)\}$ is a quasi-convex function.

Hence, in principle it can be solved by any existing algorithm for such problems, for example by (variants of) Veinott's method (see Appendix E). To apply this method we need to be able to compute $G(x)$ and its gradient ∇G. In this section we discuss how this can be done numerically if *the matrix T is fixed and $h(\omega) = \omega \in \mathbb{R}^m$ follows a multivariate normal distribution*, which is an important special case.

First of all, we claim that to perform the required calculations it is sufficient if we can evaluate the distribution function $F(z)$ of ω for arbitrary $z \in \mathbb{R}^m$,

$$
\begin{aligned}
F(z) &:= \Pr\{\omega \le z\} \\
&= \int_{-\infty}^{z_1} \cdots \int_{-\infty}^{z_m} f(t_1, \dots, t_m)\, dt_1 \dots dt_m
\end{aligned}
$$

where f is the probability density function of the m-dimensional multivariate normal distribution with mean μ and covariance matrix V, denoted by $N_m(\mu, V)$.

Indeed, writing $G(x) = \alpha - F(Tx)$ with $Tx \in \mathbb{R}^m$ this is trivially true for the evaluation of G. To prove the claim for the calculation of ∇G, we apply the chain rule to obtain

$$
\frac{\partial G}{\partial x_j}(x) = -\sum_{i=1}^{n} \frac{\partial F}{\partial z_i}(Tx) \cdot \frac{\partial (Tx)_i}{\partial x_j}
$$

or

$$
\nabla G(x) = -\nabla F(Tx) \cdot T
$$

Next we use that e.g.

$$
\begin{aligned}
\frac{\partial F}{\partial z_1}(z) &= \int_{-\infty}^{z_2} \cdots \int_{-\infty}^{z_m} f(z_1, t_2, \dots, t_m)\, dt_2 \dots dt_m \\
&= \Pr\{\omega \le z \mid \omega_1 = z_1\}.
\end{aligned}
$$

That is, the partial derivative of F with respect to z_i equals the conditional distribution of ω given $\omega_i = z_i$. The result now follows, since for multivariate normal distributions these conditional distributions are again normal: for $i = 1, \dots, m$, $\omega|\omega_i = z_i$ follows an $(m-1)$-dimensional normal distribution with known mean and variance depending on μ, V and z_i. Thus, it remains to calculate $F(z)$ for arbitrary $z \in \mathbb{R}^m$.

5.4.1 Evaluation of the Distribution Function of a Multivariate Normal Distribution

Below we review three methods to perform this task. Only the first is tailored to the normal distribution; the other two methods can also be used to evaluate the distribution function of other distributions.

First we attempt to reduce the calculation of $F(z)$ to multiple calculations of one-dimensional standard normal probabilities, which would allow to perform this task with arbitrary precision at relatively small effort. For illustrative purposes, we proceed without making any assumptions about the distribution of ω. As we will see, this allows to point out precisely why this approach only works if the ω_i are independent (which is not surprising on second thought).

Since the covariance matrix V is positive definite, there exists a regular matrix C such that $CC' = V$ (C can be determined by Choleski decomposition). Hence $\omega \equiv C\eta + \mu$ with η following an $N_m(0, I_m)$ distribution, that is, the $\eta_i, i = 1, \ldots, m$ are independent and follow a standard normal distribution ($\eta_i \sim N_1(0, 1)$ with distribution function Φ). Hence

$$\begin{aligned} F(z) &= \Pr\{\omega \leq z\} \\ &= \Pr\{C\eta + \mu \leq z\} \\ &= \Pr\{C\eta \leq z - \mu\}. \end{aligned}$$

It is tempting to continue with

$$\begin{aligned} F(z) &= \Pr\left\{\eta \leq C^{-1}(z - \mu)\right\} \\ &= \prod_{i=1}^{m} \Pr\left\{\eta_i \leq (C^{-1}(z - \mu))_i\right\} \\ &= \prod_{i=1}^{m} \Phi(c_i(z - \mu)) \end{aligned}$$

with c_i the ith row of C^{-1}. However, the first equality above is valid only if C is a diagonal matrix with strictly positive diagonal elements, which is equivalent to independence (and non-degeneracy) of the elements of ω.

Exercise 5.4.1 Give a simple proof of the result above for the case that the elements of ω are independent. ◁

Remark 5.4.2 The exposition above illustrates the fact that in general matrix multiplication does not go together with inequalities.

The conclusion is: the calculation of the value of a multivariate normal distribution function $F(z)$ by means of multiple calculations of one-dimensional standard normal distribution values is only possible if its components are stochastically independent.
The *second* possibility is to use *simulation*. Let $\omega^1, \ldots, \omega^N$ be a sample from the distribution $N_m(\mu, V)$ (which can for example be constructed using $N \cdot m$ independent samples from $N(0, 1)$. How?). Then

$$\hat{F}(z) := \frac{1}{N} \sum_{t=1}^{N} \mathbf{1}_{z+\mathbb{R}^m_-}(\omega^t),$$

where the set $z + \mathbb{R}^m_- = \{s \in \mathbb{R}^m : s_i \le z_i,\ i = 1, \ldots, m\}$ and

$$\mathbf{1}_S(\omega) = \begin{cases} 0, & \omega \notin S \\ 1, & \omega \in S, \end{cases}$$

is an estimator of $F(z)$. However, to obtain high accuracy a very large sample is needed; the confidence interval for this estimator decreases proportional to $\sqrt{N}^{-1}$. A more subtle approach, which is due to Szántai (see [33]), uses a combination of exact calculations (based on one and two-dimensional normal distributions) and simulation. This is the *third* method to be discussed. To present the main ideas of this method, we start with defining the following *simple events*: for $i = 1, \ldots, m$,

$$\begin{aligned} B_i &= \{\omega_i \le z_i\} \qquad \text{`}i\text{ is good'} \\ \overline{B}_i &= \{\omega_i > z_i\} \qquad \text{`}i\text{ is bad'} \end{aligned}$$

where the interpretations 'good' and 'bad' obviously are inspired by the fact that we are interested in $F(z) = \Pr\{\omega_i \le z_i \forall i\}$. Now $F(z)$ can be expressed as the probability of the *compound event* $\cap_{i=1}^m B_i$:

$$\begin{aligned} F(z) &= \Pr\left\{\cap_{i=1}^m B_i\right\} \\ &= \Pr\{\text{`all } i \text{ are good'}\} \\ &= 1 - \Pr\{\text{`at least one } i \text{ is bad'}\} \\ &= 1 - \Pr\left\{\cup_{i=1}^m \overline{B}_i\right\} \\ &= 1 - S_1 + S_2 - \cdots + (-1)^m S_m, \end{aligned} \tag{5.3}$$

where

$$S_k := \sum_{I \subset \{1,\ldots,m\}} \left\{\Pr\left\{\cap_{i\in I} \overline{B}_i\right\} : |I| = k\right\}$$

and the last equality follows from the *inclusion-exclusion principle*: the number of elements in a finite set that possess at least one out of m properties is equal to the number possessing exactly one of the properties (corresponding to S_1), less the number possessing exactly two properties (S_2), plus the number possessing exactly three properties (S_3), and so on, up to the number possessing all m (S_m). (See Fig. 5.3.) Thus, the computation of $F(z)$ is reduced to the computation of the probabilities S_k. The idea is now to compute the first few terms S_k explicitly, and to

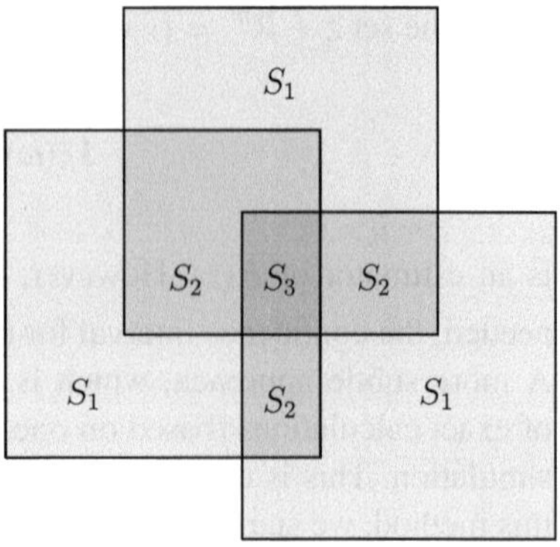

Fig. 5.3 Illustration of the inclusion-exclusion principle

use simulation to estimate the remaining part. In Szántai's method

$$S_1 = \sum_{i=1}^{m} \Pr\left\{\overline{B}_i\right\} = \sum_{i=1}^{m} \Pr\left\{\omega_i > z_i\right\}$$

$$S_2 = \sum_{i=1}^{m} \sum_{j=i+1}^{m} \Pr\left\{\omega_i > z_i, \omega_j > z_j\right\}$$

are computed explicitly.

Although the sequence $\{S_k\}$ is not monotonic in general, it follows from the *Bonferroni inequalities* that, for $n \in \{1, 2, \ldots\}$, $n < m/2$, $S_{2n} - S_{2n+1} + \cdots + (-1)^m S_m \geq 0$ and $-S_{2n+1} + \cdots + (-1)^m S_m \leq 0$. Hence, if the series (5.3) is broken off after a few terms, the result is an *upper bound* for $F(z)$ if the last term included has a plus sign; if the last term has a minus sign, the result is a *lower bound*.

It remains to construct an estimator for the tail of the series (5.3). Following Prékopa [33], we first show that S_k is the kth *binomial moment* of the random variable ξ:

$$S_k = E\left[\binom{\xi}{k}\right],$$

where ξ is the number of 'errors' in $i = 1, \ldots, m$, i.e.,

$$\xi = \sum_{i=1}^{m} \xi_i \qquad \text{with } \xi_i = \mathbf{1}_{(z_i,\infty)}(\omega_i).$$

To see that this is true, we use that

$$\binom{\xi}{k} = \sum_{\substack{k_1 \geq 0, \ldots, k_m \geq 0 \\ k_1 + \cdots + k_m = k}} \binom{\xi_1}{k_1}\binom{\xi_2}{k_2}\cdots\binom{\xi_m}{k_m}$$

$$= \sum_{|I|=k} \left\{ \prod_{i \in I} \xi_i \right\},$$

where the first equality is a well-known formula of Cauchy for binomial coefficients, and the second equality follows because, since $\xi_i \in \{0, 1\}$, $\binom{\xi_i}{0} = 1$, $\binom{\xi_i}{1} = \xi_i$, and $\binom{\xi_i}{k_i} = 0$ if $k_i \geq 2$. We see that a non-trivial contribution to this expression is obtained only if $k_i = 1$, from which there are k. We denote these, in increasing order, as $i_1, i_2, \ldots, i_k$, and we may write

$$\binom{\xi}{k} = \sum_{1 \leq i_1 < i_2 < \cdots < i_k \leq m} \xi_{i_1}\xi_{i_2}\ldots\xi_{i_k}.$$

Taking expectations, we obtain the desired result:

$$E\left[\binom{\xi}{k}\right] = \sum_{1 \leq i_1 < i_2 < \cdots < i_k \leq m} E\left[\xi_{i_1}\xi_{i_2}\ldots\xi_{i_k}\right]$$

$$= \sum_{1 \leq i_1 < i_2 < \cdots < i_k \leq m} \Pr\left\{\xi_{i_1} = 1, \xi_{i_2} = 1, \ldots, \xi_{i_k} = 1\right\}$$

$$= \sum_{\substack{I \subset \{1,\ldots,m\} \\ |I|=k}} \left\{\Pr\left\{\cap_{i \in I} \overline{B}_i\right\} : |I| = k\right\} = S_k,$$

where we used that $\xi_{i_1}\xi_{i_2}\ldots\xi_{i_k} \in \{0, 1\}$ to obtain the second equality.

Now consider an estimator for S_k, based on one observation of ω. Let ξ be the corresponding number of 'bad' indices. Then there are $\binom{\xi}{k}$ out of $\binom{m}{k}$ combinations consisting of 'bad' indices. Hence, an estimator for S_k is

$$\hat{S}_k = \sum_{\substack{I \subset \{1,\ldots,m\} \\ |I|=k}} \frac{\binom{\xi}{k}}{\binom{m}{k}} = \binom{\xi}{k}. \tag{5.4}$$

Recall that

$$F(z) = \sum_{k=0}^{h-1}(-1)^k S_k + \sum_{k=h}^{m}(-1)^k S_k,$$

and that we assume that the first h terms will be calculated analytically, and that the remaining terms are to be estimated. Using (5.4), we define

$$\tau_h(\xi) := \sum_{k=h}^{m}(-1)^k \binom{\xi}{k}.$$

It is easy to see that $\tau_h(\xi) = 0$ if $\xi < h$, and if $\xi \geq h$ then

$$\begin{aligned}
\tau_h(\xi) &= \sum_{k=h}^{\xi}(-1)^k \binom{\xi}{k} \\
&= -\sum_{k=0}^{h-1}(-1)^k \binom{\xi}{k} \qquad \text{since } \sum_{k=0}^{\xi}(-1)^k \binom{\xi}{k} = (-1+1)^{\xi} = 0 \\
&= \binom{\xi - 1}{h - 1}(-1)^h \qquad \text{since } \sum_{m=0}^{n}(-1)^{n-m}\binom{r}{m} = \binom{r-1}{m},\ r \geq n+1
\end{aligned}$$

For example,

$$\tau_1(\xi) = \begin{cases} -1, \ \xi \geq 1 \\ 0, \quad \text{otherwise} \end{cases}$$

$$\tau_2(\xi) = \begin{cases} \xi - 1, \ \xi \geq 2 \\ 0, \qquad \text{otherwise} \end{cases}$$

$$\tau_3(\xi) = \begin{cases} -\binom{\xi - 1}{2}, \ \xi \geq 3 \\ 0, \qquad \text{otherwise} \end{cases}$$

Now we are ready to define our estimator for the tail terms $\sum_{k=h}^{m}(-1)^k S_k$: based on a sample $\omega^1, \ldots, \omega^N$ it is

$$\hat{\tau}_h := \frac{1}{N}\sum_{t=1}^{N} \tau_h(\xi^t),$$

where ξ^t is the number of 'bad' indices if $\omega = \omega^t$.

Chapter 6
Integrated Chance Constraints

In this chapter we discuss some differences as well as similarities between chance-constrained and recourse models. Mixing the ideas of both model types, the concept of integrated chance constraints is introduced.

6.1 Modeling with Integrated Chance Constraints

There is a difference between chance-constrained and recourse models which seems trivial but which nevertheless is often overlooked: they use a different measure for risk. In chance constraints risk is measured *qualitatively* whereas in recourse models this is done *quantitatively*. That is, in chance constraints only the possibility of infeasibility is at stake regardless of the amount by which the constraints are violated, whereas in recourse models these amounts themselves are important. In practice it might be acceptable to have a constraint violated, perhaps even with high probability if necessary, if the amount by which it is violated is small enough. In such cases the quantitative approach of infeasibility of recourse models is more appropriate. This does not mean, however, that at the same time one has to adopt the penalty cost structure of recourse models. In certain circumstances it might be more appropriate to fix an upper bound on the risk, as in chance constraints, rather than to penalize it. This is true in particular if it is very difficult to specify the costs of infeasibility. Mixing the ideas of chance-constrained and recourse models in the above sense, we introduce the constraints

$$\mathbb{E}_{\omega}\left[\eta_i(x,\omega)^-\right] \leq \beta_i, \quad i = 1, \ldots, m, \tag{6.1}$$

W. K. Klein Haneveld et al., *Stochastic Programming*, Graduate Texts in Operations Research, https://doi.org/10.1007/978-3-030-29219-5_6

where $\eta_i(x, \omega) := T_i(\omega)x - h_i(\omega)$, and where β_i are positive risk aversion parameters to be specified in advance. Following [16], we call the constraints in (6.1) *integrated chance constraints* (ICC), since

$$\mathbb{E}_\omega \left[\eta_i(x, \omega)^-\right] = \int_{-\infty}^{0} \Pr\{\eta_i(x, \omega) < t\}\, dt.$$

Exercise 6.1.1 Prove this equality. (Hint: See the proof of the expression (3.21) for the expected shortage function H in Sect. 3.3.3.) ◁

That is, (6.1) can be seen as the integrated version of a set of chance constraints for one individual goal constraint, as described in Remark 5.1.4:

$$\Pr\{\eta_i(x, \omega) < -d_{ih}\} \le \alpha_{ih}, \quad h = 1, \ldots, H_i,$$

where $0 = d_{i1} < d_{i2} < \ldots$ and $1 > \alpha_{i1} > \alpha_{i2} > \cdots > 0$. This set of H_i chance constraints can be used instead of one individual chance constraint $\Pr\{\eta_i(x, \omega) < 0\} \le \alpha_{i0}$, to specify that large shortages are disliked more than small ones.

Remark 6.1.2 Note that in this chapter we use a different notation for (individual) chance constraints. In previous parts we used $\Pr\{\eta_i(x, \omega) \ge 0\} \ge \bar{\alpha}_i$, so that $\bar{\alpha}_i$ could be interpreted as a *reliability coefficient*. Here we formulate the constraint in terms of the complementary probability, that is, as $\Pr\{\eta_i(x, \omega) < 0\} \le 1 - \bar{\alpha}_i =: \alpha_i$, so that α_i can be interpreted as a *risk aversion parameter* comparable to the parameter β_i in (6.1).

It appears that the mathematical properties of integrated chance constraints are nicer than those of the usual chance constraints of Chap. 5. This can be explained as follows. Both $\Pr\{\eta_i(x, \omega) < 0\}$ and $\mathbb{E}_\omega\left[\eta_i(x, \omega)^-\right]$ can be written as an *expected loss function* $\mathbb{E}_\omega\left[l(\eta_i(x, \omega))\right]$; in the former case the loss function l is defined by

$$l(z) = \begin{cases} 1, & z < 0, \\ 0, & z \ge 0, \end{cases} \tag{6.2}$$

and in the latter case by

$$l(z) = \begin{cases} -z, & z < 0, \\ 0, & z \ge 0. \end{cases} \tag{6.3}$$

Comparison of both loss functions illustrates not only the difference between the 'qualitative' and the 'quantitative' measurement of risk, but it also reveals the different mathematical properties. Although both loss functions are nonincreasing, the former decreases strictly in $z_i = 0$ only, whereas the latter decreases strictly on the entire negative halfline. Moreover, the latter is convex and continuous, whereas the former is not.

It is interesting to notice that *expected loss constraints* can be interpreted as *weighted integrated chance constraints*. In order to see this, let μ be a (not necessarily finite) measure on the Borel sets of $(-\infty, 0]$ and define $l(z) := \mu((z, 0])$, $z \in \mathbb{R}$. Then, with π the probability measure on the Borel sets of $\mathbb{R}$ corresponding to the distribution of $\eta_i(x, \cdot)$, we conclude from Fubini's theorem that

$$\int_{(-\infty,0]} \left\{ \int_{(z,0]} \mu(dt) \right\} \pi(dz) = \int_{(-\infty,0]} \left\{ \int_{(-\infty,t)} \pi(dz) \right\} \mu(dt).$$

But this is nothing other than

$$\mathbb{E}_\omega \left[l(\eta_i(x, \omega)) \right] = \int_{(-\infty,0]} \Pr\{\eta_i(x, \omega) < t\}\, \mu(dt).$$

This formula shows that the use of a loss function is equivalent to a weighting of the risks $\Pr\{\eta_i(x, \omega) < t\}$. In (6.2) all the weight is concentrated in $t = 0$; in (6.3) the weights are uniform on $(-\infty, 0]$.

6.2 Single Integrated Chance Constraint

We now consider properties of a single integrated chance constraint, used to model the goal constraint $T_i(\omega)x \geq h_i(\omega)$; for convenience, we drop the index i.

6.2.1 *Formulations of Single Integrated Chance Constraints*

In view of Remark 6.1.2, we define the feasible set determined by the chance constraint $\Pr\{\eta(x, \omega) < 0\} \leq \alpha$ as

$$C_0(\alpha) := \left\{ x \in \mathbb{R}^n : \mathbb{E}_\omega \left[\operatorname{sgn}(\eta(x, \omega)^-) \right] \leq \alpha \right\}, \quad \alpha \in [0, 1], \tag{6.4}$$

where $\operatorname{sgn}(z) = 1$ if $z > 0$ and $\operatorname{sgn}(z) = 0$ otherwise. Notice that $\mathbb{E}_\omega \left[\operatorname{sgn}(\eta(x, \omega)^-) \right] = \Pr\{\eta(x, \omega) < 0\}$ so that

$$x \in C_0(\alpha) \Longleftrightarrow \Pr\{T(\omega)x \geq h(\omega)\} \geq 1 - \alpha.$$

The risk aversion parameter α denotes the maximal acceptable risk level; it has to be specified in advance. Notice that for this specification of risk the *amount of shortage* is not relevant; only its sign counts. In situations where this amount is important, it might be more appropriate to use the *mean shortage* $\mathbb{E}_\omega\left[\eta(x,\omega)^-\right]$ as a measure for risk. This leads to the following analogue of (6.4):

$$C_1(\beta) := \left\{x \in \mathbb{R}^n : \mathbb{E}_\omega\left[\eta(x,\omega)^-\right] \leq \beta\right\}, \quad \beta \in [0,\infty], \tag{6.5}$$

where the risk aversion parameter β has to be chosen in advance.

Because of (6.1) we call (6.5), and other constraints upper bounding $\mathbb{E}_\omega\left[\eta(x,\omega)^-\right]$, *integrated chance constraints* (ICCs). Whereas α in (6.4) is scale-free, β in (6.5) depends on scale. If the 'demand' h is fixed and positive, one may choose e.g. $\beta = \alpha h$ for a scale-free $\alpha \in [0,1]$. An alternative is presented in Assignment C2.

It is also possible to specify the maximum acceptable risk not as a fixed number β, but depending on the distribution of $\eta(x,\omega)$. For example, since a natural upper bound for $\mathbb{E}_\omega\left[\eta(x,\omega)^-\right]$ is given by $\mathbb{E}_\omega\left[|\eta(x,\omega)|\right]$, we introduce, for $\alpha \in [0,1]$ to be specified in advance,

$$C_2(\alpha) := \left\{x \in \mathbb{R}^n : \mathbb{E}_\omega\left[\eta(x,\omega)^-\right] \leq \alpha \cdot \mathbb{E}_\omega\left[|\eta(x,\omega)|\right]\right\}. \tag{6.6}$$

In (6.6) also the mean surplus $\mathbb{E}_\omega\left[\eta(x,\omega)^+\right]$ is taken into account, since $\mathbb{E}_\omega\left[\eta(x,\omega)^-\right] + \mathbb{E}_\omega\left[\eta(x,\omega)^+\right] = \mathbb{E}_\omega\left[|\eta(x,\omega)|\right]$. Just like (6.5) also (6.6) is a direct analogue of (6.4), at least if $\Pr\{\eta(x,\omega) = 0\} = 0$, as can be seen by rewriting the chance constraint as

$$\mathbb{E}_\omega\left[\operatorname{sgn}(\eta(x,\omega))^-\right] \leq \alpha \cdot \mathbb{E}_\omega\left[\operatorname{sgn}(|\eta(x,\omega)|)\right].$$

Obviously, each of the feasibility sets $C_0(\alpha)$, $C_1(\beta)$ and $C_2(\alpha)$ is nondecreasing in the risk aversion parameters α, β and α, respectively. They coincide for extreme specifications of these parameters:

$$C_0(0) = C_1(0) = C_2(0) = \left\{x \in \mathbb{R}^n : \Pr\{\eta(x,\omega) < 0\} = 0\right\},$$

(but, as a matter of fact, this *extremely safe set* is often empty, unfortunately), and, of course,

$$C_0(1) = C_1(\infty) = C_2(1) = \mathbb{R}^n.$$

6.2.2 Examples of Single Integrated Chance Constraints

From the previous section we know that feasibility sets $C_0(\alpha)$, $C_1(\beta)$ and $C_2(\alpha)$ coincide for extreme specifications of the risk aversion parameters α, β, and α, respectively. However, for intermediate values of the risk parameters they may behave quite differently as the following examples show. The reduced forms for the sets $C_1(\beta)$ and $C_2(\alpha)$ in these small examples can be derived easily. General results on reduced forms for these sets are presented in Sect. 6.2.3.

Example 6.2.1 Let $T(\omega) = \omega$ and $h(\omega) = 1$ with $\Pr\{\omega = -1\} = \Pr\{\omega = 1\} = 1/2$. Consider the goal constraint $\eta(x, \omega) := \omega x - 1 \geq 0$. Then, the feasibility sets unequal to $\emptyset$ or $\mathbb{R}$ are:

$$\begin{aligned} C_0(\alpha) &= \{|x| \geq 1\}, & \alpha &\in [1/2, 1), \\ C_1(\beta) &= \{|x| \leq 2\beta - 1\}, & \beta &\in [1, \infty), \\ C_2(\alpha) &= \left\{|x| \geq (2\alpha - 1)^{-1}\right\}, & \alpha &\in (1/2, 1). \end{aligned}$$

◇

Example 6.2.2 Let $T(\omega) = \omega$ and $h(\omega) = 0$, with $\Pr\{\omega = -1\} = 2/3$, $\Pr\{\omega = 1\} = 1/3$. Consider the goal constraint $\eta(x, \omega) := \omega x \geq 0$. Then, the feasibility sets are given by

$$C_0(\alpha) = C_2(\alpha) = \begin{cases} \{0\}, & 0 \leq \alpha < 1/3, \\ (-\infty, 0], & 1/3 \leq \alpha < 2/3, \\ \mathbb{R}, & 2/3 \leq \alpha \leq 1, \end{cases}$$

$$C_1(\beta) = [-3\beta, 3\beta/2], \quad 0 \leq \beta \leq \infty.$$

◇

Example 6.2.3 Let $T(\omega) = (\omega_1, \omega_2)$ and $h(\omega) = 0$, with $\Pr\{\omega = (-2, 1)\} = \Pr\{\omega = (1, -2)\} = 1/2$. Consider the goal constraint $\eta(x, \omega) := \omega_1 x_1 + \omega_2 x_2 \geq 0$. Then, the feasibility sets are given by

$$C_0(\alpha) = \begin{cases} \left\{x \in \mathbb{R}^2 : 2x_1 - x_2 \leq 0,\ 2x_2 - x_1 \leq 0\right\}, & 0 \leq \alpha < 1/2, \\ \left\{x \in \mathbb{R}^2 : 2x_1 - x_2 \leq 0\right\} \cup \left\{x \in \mathbb{R}^2 : 2x_2 - x_1 \leq 0\right\}, & 1/2 \leq \alpha < 1, \end{cases}$$

$$C_1(\beta) = \left\{ x \in \mathbb{R}^2 : \begin{array}{c} 2x_1 - x_2 \leq 2\beta, \\ 2x_2 - x_1 \leq 2\beta, \\ x_1 + x_2 \leq 2\beta \end{array} \right\}, \quad 0 \leq \beta \leq \infty,$$

$$C_2(\alpha) = \begin{cases} \left\{ x \in \mathbb{R}^2 : \begin{array}{c} (2 - 3\alpha)x_1 + (3\alpha - 1)x_2 \leq 0, \\ (2 - 3\alpha)x_2 + (3\alpha - 1)x_1 \leq 0 \end{array} \right\}, & 0 \leq \alpha \leq 1/2, \\ \left\{x \in \mathbb{R}^2 : (2 - 3\alpha)x_1 + (3\alpha - 1)x_2 \leq 0\right\} \cup & \\ \quad \left\{x \in \mathbb{R}^2 : (2 - 3\alpha)x_2 + (3\alpha - 1)x_1 \leq 0\right\}, & 1/2 \leq \alpha \leq 1. \end{cases}$$

The feasibility sets $C_0(\alpha)$ and $C_1(\beta)$ are depicted in Fig. 6.1 for various values of α and β. ◇

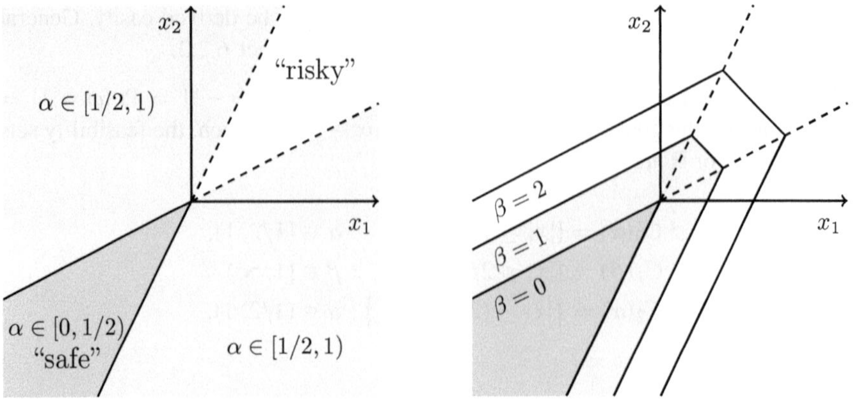

Fig. 6.1 The feasibility sets $C_0(\alpha)$ (left) and $C_1(\beta)$ (right) of Example 6.2.3

The examples illustrate that the feasibility sets determined by chance constraints have some unpleasant features, already observed in Sect. 5.2:

(i) $C_0(\alpha)$ may be *nonconvex*, even for small values of α.
(ii) $C_0(\alpha)$ may *increase step-wise* with α.

Of course, our examples are biased, in the sense that the latter difficulty does not occur if ω has a continuous distribution, and that the former difficulty disappears under certain (rather restrictive) conditions on the distribution of $(T(\omega), h(\omega))$, see Sect. 5.2.

On the other hand, the examples suggest that $C_1(\beta)$ does 'better' than $C_0(\alpha)$. That is, even for discrete distributions $C_1(\beta)$ appeared to be a convex set for all β, which increases smoothly from the extremely safe set to $\mathbb{R}^n$ if β increases from 0 to $+\infty$. It is not surprising that $C_1(\beta)$ behaves 'better' than its CC companion $C_0(\alpha)$: recall that the constraint in $C_1(\beta)$ as well as that in $C_0(\alpha)$ can be written as an expected loss constraint, where the ICC loss function

$$l_1(z) = z^-, \quad z \in \mathbb{R},$$

is convex, continuous, and strictly decreasing on $(-\infty, 0)$, whereas the CC loss function

$$l_0(z) = \operatorname{sgn}(z^-), \quad z \in \mathbb{R},$$

does not possess these properties.

The examples suggest also that $C_2(\alpha)$ behaves 'better' than $C_0(\alpha)$: it increases smoothly with α (except in Example 6.2.2) and it is convex for $\alpha \leq 1/2$, in spite of the discrete distributions. There is a basic difference between $C_2(\alpha)$ and $C_1(\beta)$. Let us call for a moment

$$R := \left\{x \in \mathbb{R}^n : \Pr\{\eta(x, \omega) < 0\} = 1\right\}$$

the *extremely risky set*, since for each $x \in R$ the constraint $T(\omega)x \geq h(\omega)$ is violated with probability 1. It is easily seen that such an extremely risky decision is not accepted as feasible by the chance constraint, as well as by the second integrated chance constraint:

$$C_0(\alpha) \cap R = C_2(\alpha) \cap R = \emptyset \quad \forall \alpha < 1.$$

However, it is quite possible that $C_1(\beta) \cap R \neq \emptyset$, even for small values of β, see e.g. Fig. 6.1. This illustrates that some values for x are accepted as feasible by the integrated chance constraint (6.5) even if it is sure that shortage will occur; the reason is, of course, that the mean shortage is small then.

6.2.3 *Properties of Single Integrated Chance Constraints*

In the previous section we discussed some examples to illustrate the behavior of the feasibility sets of chance constraints and integrated chance constraints. In this section we analyze the properties of the feasibility sets $C_1(\beta)$ and $C_2(\alpha)$ of integrated CC. As usual we assume that $\mathbb{E}_\omega[|\omega|] < \infty$. Moreover, an uninteresting case is avoided by the assumption

$$\Pr\{T(\omega) = 0\} < 1. \tag{6.7}$$

If (6.7) does not hold then the support $\mathcal{S}$ of the random vector $(T(\omega), h(\omega))$ is contained in the h-axis in $\mathbb{R}^{n+1}$. Sometimes it will appear to be appropriate to use the stronger condition

$$\mathcal{S} \text{ is not a subset of } \textit{any} \text{ line in } \mathbb{R}^{n+1} \text{ through the origin.} \tag{6.8}$$

Obviously, $C_1(\beta)$ is nondecreasing with β. For small β the set $C_1(\beta)$ may be empty: with

$$\beta_0 := \inf_{x \in \mathbb{R}^n} \mathbb{E}_\omega\left[\eta(x, \omega)^-\right], \tag{6.9}$$

which is a finite number, we have $C_1(\beta) \neq \emptyset$ if and only if

$$\begin{array}{ll} \text{either } \beta > \beta_0 & \\ \text{or} \quad \beta = \beta_0 \text{ and the infimum in (6.9) is attained.} \end{array} \tag{6.10}$$

On the other side of the range one has $C_1(\infty) = \mathbb{R}^n$. In the next theorem a general statement on the increase of $C_1(\beta)$ with β will be shown. We use the following terminology. Any family $\{C(t),\ t \in T\}$ of closed subsets $X(t) \subset \mathbb{R}^n$ where the parameter t takes values in $T := [t_0, t_1] \subset [0, \infty]$ is called *nondecreasing* if

$$C(s) \subset C(t) \text{ if } s \in T,\ t \in T,\ s < t,$$

and it is called *strictly increasing* at $t \in T$ if it is nondecreasing at $t \in T$ and if

$$C(s) \neq C(t) \text{ for all } s \in T,\ s \neq t;$$

and it is called *continuously increasing* at $t \in T$ if it is nondecreasing at $t \in T$ and if

$$\operatorname{cl} \bigcup_{s \in [t_0, t)} C(s) = C(t) = \bigcap_{s \in (t, t_1]} C(s).$$

Here it is understood that for $t = t_0$ only the second condition has to hold, and for $t = t_1$ only the first one.

Theorem 6.2.4 *Assume that* $\mathbb{E}_\omega [|\omega|] < \infty$.

(a) For any $\beta \in [0, \infty]$, $C_1(\beta)$ *is a closed, convex subset of* $\mathbb{R}^n$. *It is nonempty if and only if (6.10) holds, and* $C_1(\beta) \neq \mathbb{R}^n$ *for all* $\beta < \infty$ *if (6.7) holds.*
(b) $C_1(\beta)$ *is nondecreasing with* β. *The increase is continuous at all* $\beta \neq \beta_0$; *and it is strict at all* $\beta > \beta_0$ *if (6.7) holds.*
(c) Let $\beta < \infty$ *with* $C_1(\beta) \neq \emptyset$. *Then*

$$\left.\begin{array}{l} x + \lambda y \in C_1(\beta) \\ \forall x \in C_1(\beta),\ \forall \lambda \geq 0 \end{array}\right\} \Longleftrightarrow \Pr\{T(\omega) y \geq 0\} = 1.$$

Therefore, $C_1(\beta)$ *is bounded if and only if* $0 \in \operatorname{int} \operatorname{conv} \mathcal{T}$, *where* $\mathcal{T} \subset \mathbb{R}^n$ *is the support of the distribution of* $T(\omega) = (t_1(\omega), \ldots, t_n(\omega))$.

Proof For the proof we refer to [16]. □

We will now present some properties of the second type of ICC, introduced in (6.6).

Since for all $z \in \mathbb{R}$ it holds that $z = (z)^+ - (z)^-$ and $|z| = (z)^+ + (z)^-$, we have

$$\begin{aligned} C_2(\alpha) &= \left\{x \in \mathbb{R}^n : \mathbb{E}_\omega \left[\eta(x, \omega)^-\right] \leq \alpha \cdot \mathbb{E}_\omega \left[|\eta(x, \omega)|\right]\right\} \\ &= \left\{x \in \mathbb{R}^n : (1 - \alpha) \cdot \mathbb{E}_\omega \left[\eta(x, \omega)^-\right] \leq \alpha \cdot \mathbb{E}_\omega \left[\eta(x, \omega)^+\right]\right\} \qquad (6.11) \\ &= \left\{x \in \mathbb{R}^n : (1 - 2\alpha) \cdot \mathbb{E}_\omega \left[\eta(x, \omega)^-\right] \leq \alpha \cdot \mathbb{E}_\omega \left[\eta(x, \omega)\right]\right\}. \end{aligned}$$

The representation (6.11) shows that in (6.6) a positive mean shortage is only accepted if the corresponding mean surplus is large enough.

Theorem 6.2.5 *Assume that $\mathbb{E}_\omega[|\omega|] < \infty$ and that (6.8) holds. Then*

(a) For any $\alpha \in [0, 1]$ $C_2(\alpha)$ is a closed subset of $\mathbb{R}^n$.

(b) If $\alpha \leq 1/2$ then $C_2(\alpha)$ is convex; it is even polyhedral if $(T(\omega), h(\omega))$ has a finite discrete distribution. If $\alpha > 1/2$ then $C_2(\alpha)$ is the complement in $\mathbb{R}^n$ of a convex set, hence nonconvex generally. For $\alpha = 1/2$ the integrated chance constraint in (6.6) has a simple deterministic equivalent representation in terms of the expected values of the coefficients:

$$C_2(1/2) = \left\{x \in \mathbb{R}^n : \bar{T}x \geq \bar{h}\right\},$$

where $(\bar{T}, \bar{h})$ is the mean value of $(T(\omega), h(\omega))$.

If $\Pr\{h(\omega) = 0\} = 1$ then $C_2(\alpha)$ is a nonempty cone for all $\alpha \in [0, 1]$.

(c) $C_2(\alpha)$ is nondecreasing with α.

(d) For $\alpha \in (1/2, 1]$ $C_2(\alpha)$ is nonempty and unbounded. For $\alpha \in [0, 1/2]$ with $C_2(\alpha) \neq \emptyset$

$$\left.\begin{array}{l} x + \lambda y \in C_2(\alpha) \\ \forall x \in C_2(\alpha), \forall \lambda \geq 0 \end{array}\right\} \Leftrightarrow \mathbb{E}_\omega\left[(T(\omega)y)^-\right] \leq \alpha \cdot \mathbb{E}_\omega[|T(\omega)y|].$$

Denote this condition on $y \in \mathbb{R}^n$ by $y \in \hat{C}_2(\alpha)$. Then any nonempty $C_2(\alpha)$, $\alpha \in [0, 1/2]$, is bounded if and only if $\hat{C}_2(\alpha) = \{0\}$. In that case, $0 \in \operatorname{int}\operatorname{conv}\mathcal{T}$, where $\mathcal{T} \subset \mathbb{R}^n$ is the support of the distribution of $T(\omega) = (t_1(\omega), \ldots, t_n(\omega))$.

Proof For the proof we refer to [16]. □

Remark 6.2.6 In addition, precise conditions are known such that $C_2(\alpha)$ is nonempty and $C_2(\alpha) \neq \mathbb{R}^n$. However, these conditions are too technical to present them here. Also, scalars α_0 and α_1 can be specified, such that $C_2(\alpha)$ increases continuously at all $\alpha \neq \alpha_0$, and strictly at all $\alpha \in (\alpha_0, \alpha_1)$. For details we refer to Klein Haneveld [16].

The choice $\alpha = 1/2$ in $C_0(\alpha)$ reflects a more or less risk-neutral attitude: the median of $\eta(x, \omega)$ must be nonnegative. Similarly, $\alpha = 1/2$ in $C_2(\alpha)$ corresponds to a risk-neutral attitude: the mean value of $\eta(x, \omega)$ must be nonnegative. In fact, the practice of neglecting randomness in the constraints may be interpreted as replacing $T(\omega) \geq h(\omega)$ by $C_2(1/2)$. Consequently, in $C_2(\alpha)$ one should specify $\alpha < 1/2$ just as in usual CCs, if one is risk-averse. In unfortunate circumstances risk is unavoidable: $C_2(\alpha) = \emptyset$ for all $\alpha \in [0, 1/2]$, see for instance Example 6.2.1.

In Sect. 5.2 we have seen that in special cases the chance constraint in the definition of $C_0(\alpha)$ has a deterministic equivalent constraint which can be calculated easily. In particular this is true if only the right-hand side is random, or if the joint distribution is normal. Similar deterministic equivalents will be derived for the integrated chance constraints related to $C_1(\beta)$ and $C_2(\alpha)$, also for the case that ω is discretely distributed.

Only RHS Random

Suppose first that $T(\omega)$ is deterministic. Then $\mathbb{E}_\omega\left[\eta(x,\omega)^-\right] = G(Tx)$ with $G(z) := \mathbb{E}_\omega\left[(z-h(\omega))^-\right]$. The properties of this function G, which is the familiar expected surplus function defined in Sect. 3.3.3, are listed in Theorem 3.3.12. From this characterization of the function G one easily derives deterministic equivalents for $C_1(\beta)$ and $C_2(\alpha)$; they have the form $\{x : Tx \geq k\}$ for a certain k, as is well-known for $C_0(\alpha)$.

Theorem 6.2.7 *Assume that $T(\omega) = T$ is fixed, $h(\omega) = \omega$ with $\mathbb{E}_\omega[\omega] = \bar{\omega}$, and that $\mathbb{E}_\omega[|\omega|] < \infty$. Denote $F(t) := \Pr\{\omega < t\}$. Then*

$$C_0(\alpha) = \left\{x \in \mathbb{R}^n : Tx \geq k_\alpha^0\right\}, \quad 0 < \alpha < 1,$$

where $k_\alpha^0 := \inf F^{-1}(1-\alpha)$.

$$C_1(\beta) = \left\{x \in \mathbb{R}^n : Tx \geq k_\beta^1\right\}, \quad 0 < \beta < \infty,$$

where k_β^1 is the unique solution of $G(k) = \beta$.

$$C_2(\alpha) = \left\{x \in \mathbb{R}^n : Tx \geq k_\alpha^2\right\}, \quad 0 < \alpha < 1,$$

where k_α^2 is the unique solution of $(1-2\alpha)G(k) = \alpha(k-\bar{\omega})$.

The functions $\beta \mapsto k_\beta^1$, $0 < \beta < \infty$, and $\alpha \mapsto k_\alpha^2$, $0 < \alpha < 1$, are continuous and strictly decreasing, and

$$\lim_{\alpha\downarrow 0} k_\alpha^0 = \lim_{\beta\downarrow 0} k_\beta^1 = \lim_{\alpha\downarrow 0} k_\alpha^2 = \sup\{t : F(t) < 1\},$$

$$\lim_{\alpha\uparrow 1} k_\alpha^0 = \lim_{\alpha\uparrow 1} k_\alpha^2 = \inf\{t : F(t) > 0\},$$

$$\lim_{\beta\uparrow\infty} \left(k_\beta^1 - (\bar{\omega} - \beta)\right) = 0.$$

Moreover, $k_{1/2}^2 = \bar{\omega}$.

Exercise 6.2.8 Prove Theorem 6.2.7. ◇

Normal Distributions

Suppose now that $(T(\omega), h(\omega))$ are normally distributed with mean $(\bar{T}, \bar{h})$ and covariance matrix V. In this case $\eta(x,\omega)$ has also a normal distribution, with mean value $\mu(x) = \bar{T}x - \bar{h}$ and variance $\sigma^2(x) = \tilde{x}'V\tilde{x}$, with $\tilde{x} := (x, -1)$. It is well-known that $\sigma(x)$ is a convex function. If $\sigma(x) = 0$ then $\mathbb{E}_\omega\left[\eta(x,\omega)^-\right] = \mu(x)^-$, and if $\sigma(x) > 0$

$$\mathbb{E}_\omega\left[\eta(x,\omega)^-\right] = \sigma(x) G_1\left(\frac{\mu(x)}{\sigma(x)}\right),$$

where $G_1(z) := \mathbb{E}_\xi\left[(z-\xi)^-\right]$, with ξ a standard normal random variable. Obviously, the analysis of G_1 follows from that of the expected (continuous) surplus function G (see Sect. 3.3.3). In particular, using that the pdf φ of ξ satisfies $\varphi'(z) = -z\varphi(z)$,

$$G_1(z) = -z + z\Phi(z) + \varphi(z), \quad z \in \mathbb{R},$$

where Φ is the cdf of the standard normal distribution.

Theorem 6.2.9 *Assume that $(T(\omega), h(\omega))$ are normally distributed. Then*

$$C_0(\alpha) = \left\{x \in \mathbb{R}^n : \mu(x) \geq \kappa^0_\alpha \cdot \sigma(x)\right\}, \quad 0 < \alpha < 1,$$

where $\kappa^0_\alpha := \Phi^{-1}(1-\alpha)$.

$$C_1(\beta) = \left\{x \in \mathbb{R}^n : \mu(x) \geq \kappa^1_{\beta/\sigma(x)} \cdot \sigma(x)\right\}, \quad 0 < \beta < \infty,$$

where κ^1_β is the unique solution of $G_1(k) = \beta$.

$$C_2(\alpha) = \left\{x \in \mathbb{R}^n : \mu(x) \geq \kappa^2_\alpha \cdot \sigma(x)\right\}, \quad 0 < \alpha < 1,$$

where κ^2_α is the unique solution of $(1-2\alpha)G_1(k) = \alpha k$.

$C_0(\alpha)$ and $C_2(\alpha)$ are convex for $\alpha \leq 1/2$; $C_1(\beta)$ is convex for all β. The functions $\alpha \mapsto \kappa^0_\alpha$, $\beta \mapsto \kappa^1_\beta$, $\alpha \mapsto \kappa^2_\alpha$, are continuous and strictly decreasing, with $\kappa^0_{1/2} = \kappa^2_{1/2} = 0$, $\kappa^1_\beta = 0$ if $\beta = (2\pi)^{-1/2}$, and

$$\lim_{\alpha\downarrow 0} \kappa^0_\alpha = \lim_{\beta\downarrow 0} \kappa^1_\beta = \lim_{\alpha\downarrow 0} \kappa^2_\alpha = +\infty,$$

$$\lim_{\alpha\uparrow 1} \kappa^0_\alpha = \lim_{\alpha\uparrow 1} \kappa^2_\alpha = -\infty,$$

$$\lim_{\beta\uparrow\infty} \left(\kappa^1_\beta + \beta\right) = 0.$$

Exercise 6.2.10 Prove Theorem 6.2.9. ◁

6.2.4 Discrete Distributions

All preceding results illustrate that tractable reduced forms of integrated chance constraints can be obtained whenever this is possible for ordinary chance constraints. The following result, obtained in [22], presents such a reduced form for the case with discrete distributions.

Theorem 6.2.11 *Assume that ω is a* discrete *random vector, with* $\Pr\{\omega = \omega^s\} = p^s$*, $s \in S$, and let $(T^s, h^s) = (T(\omega^s), h(\omega^s))$ for $s \in S$. Then, for $\beta \geq 0$,*

$$C_1(\beta) = \bigcap_{K \subset S} \left\{ x \in \mathbb{R}^n : \sum_{k \in K} p^k \left(h^k - T^k x \right) \leq \beta \right\}. \tag{6.12}$$

If S is a finite set, then $C_1(\beta)$ is a polyhedral set defined by $2^{|S|}-1$ linear constraints.

Proof Since ω is discretely distributed, we have

$$\begin{aligned} \mathbb{E}_\omega \left[\eta(x, \omega)^- \right] &= \sum_{s \in S} p^s \max\{0, -\eta(x, \omega^s)\} \\ &= \sum_{s \in S} \max\{0, -p^s \eta(x, \omega^s)\} \\ &= \sum_{s \in S} \left(-p^s \eta(x, \omega^s) \right)^+ \\ &= \max_{K \subset S} \sum_{k \in K} -p^k \eta(x, \omega^k), \end{aligned} \tag{6.13}$$

where the last equality follows from the observation that $\max_{K \subset S} \sum_{k \in K} a_k$, $a_k \in \mathbb{R}$, is attained for the subset $\{k \in S : a_k > 0\}$, see Appendix A. Moreover, for $m_i \in \mathbb{R}$, $i \in N$, with N an arbitrary index set, it holds that $\sup_{i \in N} m_i \leq M$ if and only if $m_i \leq M$ for all $i \in N$, so that

$$\begin{aligned} C_1(\beta) &= \bigcap_{K \subset S} \left\{ x \in \mathbb{R}^n : -\sum_{k \in K} p^k \eta(x, \omega^k) \leq \beta \right\} \\ &= \bigcap_{K \subset S} \left\{ x \in \mathbb{R}^n : \sum_{k \in K} p^k \left(h^k - T^k x \right) \leq \beta \right\}. \end{aligned}$$

If S is finite then there are $2^{|S|}-1$ non-empty subsets of S, so that (6.12) describes the convex set $C_1(\beta)$ using finitely many linear constraints. That is, $C_1(\beta)$ is a polyhedral set in this case. □

Remark 6.2.12 For $K = S$ in (6.12), we obtain $\bar{T}x \geq \bar{h} - \beta$, where $\bar{T}$ and $\bar{h}$ are defined as $(\bar{T}, \bar{h}) = \mathbb{E}_\omega[(T(\omega), h(\omega))]$. By our assumption that $T(\omega)$ and $h(\omega)$ depend linearly on ω, this is equivalent to $\eta(x, \mathbb{E}_\omega[\omega])^- \leq \beta$. This is an obvious necessary condition for $x \in C_1(\beta)$, since $\eta(x, \omega)^- \geq -\eta(x, \omega)$ for all ω so that $\mathbb{E}_\omega\left[\eta(x, \omega)^-\right] \geq -\eta(x, \mathbb{E}_\omega[\omega])$.

Also, the subset $K = \emptyset$ gives the trivial constraint $0 \leq \beta$, which stresses the obvious fact that $C_1(\beta) = \emptyset$ for all $\beta < 0$.

Remark 6.2.13 Theorem 6.2.11 can be generalized to a result for arbitrary distributions, as follows: for $\beta \geq 0$,

$$C_1(\beta) = \bigcap_{B \in \mathcal{B}} \left\{ x \in \mathbb{R}^n : \mathbb{E}_\omega[-\eta(x, \omega) \cdot I_B(\omega)] \leq \beta \right\}, \tag{6.14}$$

where $\mathcal{B}$ is the collection of all Borel sets in $\mathbb{R}$, and the indicator function $I_B(\omega) = 1$ if and only if $\omega \in B$.

The proof of (6.14) is omitted here.

We conclude from Theorem 6.2.11 that problems with individual ICC constraints and finite distributions can be represented as LP problems. In fact, by using discrete approximations of arbitrary (continuous) distributions, Theorem 6.2.11 implies that any ICC problem can at least be approximated by an LP problem. Note that such a discrete approximation does not work well for traditional CC problems, since in that case the approximating problem is non-convex in general. On the other hand, they fit in the framework of *mixed-integer* linear programming, as is explained in Sect. 5.3.

Finally, we present reduced forms for the sets $C_2(\alpha)$, $\alpha \in [0, 1]$, for the case that ω is discretely distributed.

Theorem 6.2.14 *Assume that ω is a* discrete *random vector, with* $\Pr\{\omega = \omega^s\} = p^s$, $s \in S$.

(i) If $\alpha \in [0, 1/2]$ then

$$C_2(\alpha) = \bigcap_{K \subset S} \left\{ x \in \mathbb{R}^n : (1 - 2\alpha) \sum_{k \in K} p^k \left(h^k - T^k x\right) \leq \alpha \left(\bar{T}x - \bar{h}\right) \right\}$$

where $(T^s, h^s) = (T(\omega^s), h(\omega^s))$, $s \in S$, and $(\bar{T}, \bar{h}) = \mathbb{E}_\omega[(T(\omega), h(\omega))]$. If S is a finite set, then $C_2(\alpha)$ is a polyhedral convex set defined by $2^{|S|} - 1$ linear constraints.

(ii) If $\alpha > 1/2$ then

$$C_2(\alpha) = \bigcup_{K \subset S} \left\{ x \in \mathbb{R}^n : (1 - 2\alpha) \sum_{k \in K} p^k \left(h^k - T^k x\right) \leq \alpha \left(\bar{T}x - \bar{h}\right) \right\}.$$

In this case $C_2(\alpha)$ is the union *of convex sets (polyhedra if S is finite), which is non-convex in general.*

Proof

(i) If $\alpha < 1/2$ (so that $1 - 2\alpha > 0$), the same approach as in the proof of Theorem 6.2.11 leads to the result. For $\alpha = 1/2$ the result is trivial.
(ii) If $\alpha > 1/2$ we find that

$$C_2(\alpha) = \left\{x \in \mathbb{R}^n : \mathbb{E}_\omega\left[\eta(x,\omega)^-\right] \geq \frac{\alpha}{1-2\alpha} \cdot \mathbb{E}_\omega\left[\eta(x,\omega)\right]\right\}.$$

Applying (6.13), and using that $\max_{i \in N} m_i \geq M$ if and only if $\exists i \in N : m_i \geq M$ (with $m_i \in \mathbb{R}$, $i \in N$; N finite), we obtain

$$\begin{aligned} C_2(\alpha) &= \bigcup_{K \subset S}\left\{x \in \mathbb{R}^n : \sum_{k \in K} -p^k \eta(x,\omega^k) \geq \frac{\alpha}{1-2\alpha}\mathbb{E}_\omega\left[\eta(x,\omega)\right]\right\} \\ &= \bigcup_{K \subset S}\left\{x \in \mathbb{R}^n : (1-2\alpha)\sum_{k \in K} p^k\left(h^k - T^k x\right) \leq \alpha\left(\bar{T}x - \bar{h}\right)\right\}. \end{aligned}$$

□

In practice, risk aversion is modeled by choosing $\alpha \ll 1/2$. As can be seen from Theorem 6.2.14, the set $C_2(\alpha)$ is convex polyhedral in this case, at least if the underlying distribution is finite.

ICC for Equality (Goal) Constraints

Up to now we considered ICC formulations for an inequality constraint $T(\omega)x \geq h(\omega)$. Whereas the CC formulation of an *equality constraint* $T(\omega)x = h(\omega)$ is useless, this is not true for ICCs. For example, one may define

$$C_4(\beta_1, \beta_2) := \left\{x \in \mathbb{R}^n : \begin{array}{l} \mathbb{E}_\omega\left[\eta(x,\omega)^-\right] \leq \beta_1, \\ \mathbb{E}_\omega\left[\eta(x,\omega)^+\right] \leq \beta_2 \end{array}\right\},$$

$$C_5(\beta) := \left\{x \in \mathbb{R}^n : \mathbb{E}_\omega\left[|\eta(x,\omega)|\right] \leq \beta\right\}.$$

Both sets are convex, but they are empty if the risk parameters are specified too small.

Exercise 6.2.15 Show this. ◁

6.3 Joint Integrated Chance Constraints

In this short section we consider m random constraints, $m \geq 2$,

$$T_i(\omega)x \geq h_i(\omega), \quad i = 1, \dots, m, \tag{6.15}$$

where $(T(\omega), h(\omega))$ has a known distribution with finite means. If for each constraint a separate degree of risk aversion is specified, (6.15) can be reformulated in terms of ICCs as

$$x \in \tilde{C}_1(\beta_1, \dots, \beta_m) := \bigcap_{i=1}^{m} C_{1i}(\beta_i), \quad 0 \leq \beta_i \leq \infty, \tag{6.16}$$

$$x \in \tilde{C}_2(\alpha_1, \dots, \alpha_m) := \bigcap_{i=1}^{m} C_{2i}(\alpha_i), \quad 0 \leq \alpha_i \leq 1, \tag{6.17}$$

where $C_{1i}(\beta_i)$ and $C_{2i}(\alpha_i)$ are defined as in (6.5) and (6.6). Of course, $\tilde{C}_1$ and $\tilde{C}_2$ have the same properties as C_1 and C_2.

A well-known alternative for separate CCs is the *joint* CC

$$D_0(\alpha) := \left\{x \in \mathbb{R}^n : \Pr\left\{T(\omega)x \geq h(\omega)\right\} \geq 1 - \alpha\right\}, \quad 0 \leq \alpha \leq 1. \tag{6.18}$$

Since $D_0(\alpha)$ can be formulated as

$$D_0(\alpha) = \left\{x \in \mathbb{R}^n : \mathbb{E}_\omega\left[\operatorname{sgn} \max_i \eta_i(x, \omega)^-\right] \leq \alpha\right\},$$

where $\eta_i(x, \omega) := T_i(\omega)x - h_i(\omega)$, the obvious generalization of $C_1(\beta)$ to joint ICC is

$$D_1(\beta) := \left\{x \in \mathbb{R}^n : \mathbb{E}_\omega\left[\max_i \eta_i(x, \omega)^-\right] \leq \beta\right\}, \quad 0 \leq \beta \leq \infty. \tag{6.19}$$

Unlike $D_0(\alpha)$, $D_1(\beta)$ is *convex* for all distributions of $(T(\omega), h(\omega))$ and for all values of the risk aversion parameter. In fact, the same statement is true for

$$D_1'(\beta) := \left\{x \in \mathbb{R}^n : \mathbb{E}_\omega\left[\|\eta(x, \omega)^-\|\right] \leq \beta\right\}, \quad 0 \leq \beta \leq \infty.$$

$$D_1''(\beta) := \left\{x \in \mathbb{R}^n : \mathbb{E}_\omega\left[l(\eta(x, \omega))\right] \leq \beta\right\}, \quad -\infty < \beta \leq \infty,$$

where $\|\cdot\|$ denotes *any* norm in $\mathbb{R}^n$, and l denotes *any* convex multi-attribute loss function, respectively. Joint variants of $C_2(\alpha)$ are less obvious. One might define e.g. for $0 \le \alpha \le 1$,

$$D_2(\alpha) := \left\{x \in \mathbb{R}^n : \mathbb{E}_\omega\left[\max_i \eta_i(x,\omega)^-\right] \le \alpha \cdot \mathbb{E}_\omega\left[\max_i |\eta_i(x,\omega)|\right]\right\},$$

$$D_2'(\alpha) := \left\{x \in \mathbb{R}^n : (1-\alpha)\cdot \mathbb{E}_\omega\left[\max_i \eta_i(x,\omega)^-\right] \le \alpha \cdot \mathbb{E}_\omega\left[\max_i \eta_i(x,\omega)^+\right]\right\},$$

$$D_2''(\alpha) := \left\{x \in \mathbb{R}^n : (1-2\alpha)\cdot \mathbb{E}_\omega\left[\max_i \eta_i(x,\omega)^-\right] \le \alpha \cdot \mathbb{E}_\omega\left[\min_i \eta_i(x,\omega)\right]\right\},$$

each of which reduces to $C_2(\alpha)$ if $m = 1$. $D_2(\alpha)$ and $D_2'(\alpha)$ are nonconvex generally, but $D_2''(\alpha)$ is convex for $\alpha \le 1/2$.

For the case that ω follows a finite discrete distribution, Klein Haneveld and Van der Vlerk [22] show that the reduced form of the sets $D_1(\beta)$, $\beta \ge 0$, is a convex polyhedral set.

6.4 Conditional Surplus-at-Risk and Integrated Chance Constraints

In this section we consider the relation between ICC constraints and *Conditional Surplus-at-Risk* (CSaR) constraints. CSaR is a variant of the Conditional Value-at-Risk concept, which is very popular, in particular in financial applications. Here we consider CSaR because it is defined in terms of surplus, similar to ICC which is based on shortage. We refer to [45] for an overview of recent results in this area.

Let the random variable ξ represent a *yield*, so that higher values are preferred. For the moment, we suppress the dependence of ξ on the decision variables x, e.g., $\xi = \eta(\omega, x)$. To avoid technicalities, we will assume that ξ has finite mean value μ and is continuously distributed, with probability density function f satisfying $f(t) > 0$ for all $t \in \mathbb{R}$, so that its cumulative distribution function F is invertible.

Definition 6.4.1 For $\gamma \in (0,1)$, the *Conditional Surplus-at-Risk* at probability level γ, associated with the random variable ξ, is

$$\text{CSaR}(\gamma) \equiv c(\gamma) := \mathbb{E}_\xi\left[\xi \mid \xi \le v(\gamma)\right],$$

where $v(\gamma)$ is the *Surplus-at-Risk* at level γ,

$$\text{SaR}(\gamma) \equiv v(\gamma) := \max_{t \in \mathbb{R}} \{t : \Pr\{\xi > t\} \ge \gamma\}.$$

Thus, Surplus-at-Risk (SaR) at level γ is defined as the $(1-\gamma)$-quantile of the distribution of ξ, and CSaR is the conditional expectation of the tail of the distribution up to SaR. It follows from our assumption on the distribution of ξ that $v(\gamma) = F^{-1}(1-\gamma)$, $\gamma \in (0,1)$.

First we show that SaR and CSaR can be found as the solution and the optimal value, respectively, of a convex optimization problem involving the expected shortage function $H(t) := \mathbb{E}_\xi\left[(\xi - t)^-\right]$, $t \in \mathbb{R}$. The convex function H is the (continuous) expected shortage function, well-known from the study of simple recourse models; its properties are presented in Theorem 3.3.10.

Lemma 6.4.2 *(i) For $\gamma \in (0, 1)$, the SaR $v(\gamma)$ satisfies*

$$v(\gamma) = \operatorname*{argmax}_{t \in \mathbb{R}} \left(t - \frac{H(t)}{1-\gamma} \right).$$

(ii) For $\gamma \in (0, 1)$, the CSaR $c(\gamma)$ satisfies

$$c(\gamma) = \max_{t \in \mathbb{R}} \left(t - \frac{H(t)}{1-\gamma} \right),$$

so that

$$c(\gamma) = v(\gamma) - \frac{H(v(\gamma))}{1-\gamma}.$$

Proof (i) Because ξ is continuously distributed, the function H is differentiable with derivative F. The first part then follows from the observation that $H'(v(\gamma)) = F(v(\gamma)) = 1 - \gamma$.

(ii) By Definition 6.4.1, $c(\gamma) = m(v(\gamma))$ with $m(t) := \mathbb{E}_\xi\,[\xi \,|\, \xi \leq t]$, $t \in \mathbb{R}$. Since, for $t \in \mathbb{R}$,

$$\begin{aligned} H(t) &= \Pr\{\xi \leq t\} \cdot \mathbb{E}_\xi\,[t - \xi \,|\, \xi \leq t] + \Pr\{\xi > t\} \cdot 0 \\ &= F(t) \cdot \left(t - \mathbb{E}_\xi\,[\xi \,|\, \xi \leq t]\right) \\ &= F(t) \cdot \left(t - m(t)\right), \end{aligned}$$

it follows that

$$m(t) = t - \frac{H(t)}{F(t)}, \qquad t \in \mathbb{R}.$$

Part (ii) now follows from (i) by substitution. □

See Fig. 6.2 for an illustration of Lemma 6.4.2.

Remark 6.4.3 Similar results, formulated in terms of the convex conjugate of the function H, have been obtained independently in [30].

In the context of decision models it is natural to assume that the random surplus depends on the decision variables $x \in \mathbb{R}^n$. In agreement with previous notation, we assume that the underlying yield ξ depends on x via $\xi = \eta(\omega, x) = T(\omega)x - h(\omega)$,

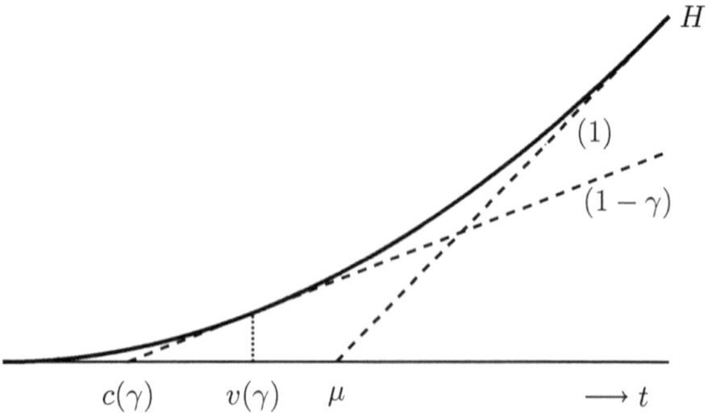

Fig. 6.2 Illustration of Lemma 6.4.2: CSaR and SaR related to the function H (solid). CSaR equals the value of t where the tangent with slope $(1 - \gamma)$ intersects the t-axis; SaR is the t-coordinate of the tangent point

where the n-vector $T(\omega)$ and the scalar $h(\omega)$ depend linearly on the underlying random vector ω. So the Conditional Surplus-at-Risk CSaR(γ) of η is now $c(\gamma, x)$. Requiring that a decision x is only accepted if its CSaR value is at least c_0, leads to the constraint

$$c(\gamma, x) \geq c_0.$$

By Lemma 6.4.2, this CSaR constraint is equivalent to

$$\mathbb{E}_\omega\left[\left(\eta(\omega, x) - t\right)^-\right] \leq (1 - \gamma)(t - c_0), \qquad t = v(\gamma, x) = \text{SaR}(\gamma, x). \quad (6.20)$$

If we choose to consider t as a parameter instead, with the interpretation of a threshold level to be met, (6.20) is just an individual ICC constraint of the first type. For any fixed t, the aspiration parameter c_0 should not be set too high, since then no feasible solutions exist. This corresponds to the fact that $v(\gamma, x)$ is a natural upper bound for c_0 in the CSaR constraint, since $c(\gamma, x) \leq v(\gamma, x)$ by definition.

6.5 Integrated Chance Constraints and Recourse Models

Just as in ICCs, in simple recourse models 'risk' is interpreted as 'mean shortage'. But risk aversion is specified in terms of penalty cost for shortages rather than by prescribing the maximal acceptable risk. It is clear, that there must be an intimate relation between both model types, expressed by Lagrange multipliers. In this section we will show that for each convex ICC model an equivalent (often simple) recourse model exists. Hence, the criticism that CC programming is a deficient modeling technique because of the poor correspondence between CC models and recourse models, is not valid for convex ICCs.

Suppose that the finite convex objective function $c(x)$ is to be minimized on the nonempty convex set X. First, we consider the *ICC described in (6.16)*. Then we get

$$\mathrm{ICC}_1(\beta) \quad \min_{x\in X}\left\{c(x) : \mathbb{E}_\omega\left[\left(T_i(\omega)x - h_i(\omega)\right)^-\right] \le \beta_i,\ i = 1,\dots,m\right\}, \quad \beta \ge 0.$$

This convex program is strictly feasible if β is not too small, that is: for all i, β_i is larger than $\beta_{i0} := \mathbb{E}_\omega\left[(T_i(\omega)\hat{x} - h_i(\omega))^-\right]$, for a suitable $\hat{x} \in X$. The Lagrangian problem

$$\mathrm{L}_1(\lambda) \quad \min_{x\in X}\left\{c(x) + \sum_{i=1}^m \lambda_i \cdot \mathbb{E}_\omega\left[\left(T_i(\omega)x - h_i(\omega)\right)^-\right]\right\}, \quad \lambda \ge 0,$$

is a simple recourse model, for which the following results hold.

Theorem 6.5.1

(a) If $\beta_i > \beta_{i0}$ $\forall i$ then there exist optimal Lagrange multipliers λ_i^0 for the constraints in $ICC_1(\beta)$, and $x^* \in X$ solves $ICC_1(\beta)$ if and only if it is feasible for it and if it is a solution of $L_1(\lambda^0)$.

(b) If $x^* \in X$ solves $\mathrm{L}_1(\lambda)$ for any $\lambda \ge 0$, then it also solves the minimization problem $ICC_1(\beta)$ with $\beta_i := \mathbb{E}_\omega\left[\left(T_i(\omega)x^* - h_i(\omega)\right)^-\right]$.

Proof Follows directly from the Karush–Kuhn–Tucker Theorem. The condition on β_i implies that Slater's constraint qualification holds. □

Of course, the mathematical equivalence in Theorem 6.5.1 does not mean that both models are equivalent from a practical point of view, since the specification of penalty costs may be more difficult or less difficult than that of risk aversion parameters β.

Consider now the separate ICCs (6.17). With obvious adjustments Theorem 6.5.1 part (a) remains true if $\mathrm{ICC}_1(\beta)$ and $\mathrm{L}_1(\lambda)$ are replaced by $\mathrm{ICC}_2(\alpha)$ and $\mathrm{L}_2(\lambda, \alpha)$, respectively: for $\alpha_i \in [0, 1/2]$

$$\mathrm{ICC}_2(\alpha) \quad \min_{x\in X}\left\{c(x) : \mathbb{E}_\omega\left[\left(T_i(\omega)x - h_i(\omega)\right)^-\right] \le \alpha_i \cdot \mathbb{E}_\omega\left[\left|T_i(\omega)x - h_i(\omega)\right|\right],\ i = 1,\dots,m\right\},$$

$$\mathrm{L}_2(\lambda, \alpha) \quad \min_{x\in X}\left\{c(x) + \sum_{i=1}^m (1-\alpha_i)\lambda_i \cdot \mathbb{E}_\omega\left[\left(T_i(\omega)x - h_i(\omega)\right)^-\right] + \sum_{i=1}^m (-\alpha_i)\lambda_i \cdot \mathbb{E}_\omega\left[\left(T_i(\omega)x - h_i(\omega)\right)^+\right]\right\}, \quad \lambda \ge 0.$$

Also for *joint integrated chance constraints (6.19)* equivalence to a recourse model can be proved. In this case the recourse is fixed and complete, but not simple. Defining

$$\text{ICC}_3(\beta) \quad \min_{x\in X} \left\{ c(x) : \mathbb{E}_\omega \left[\max_i \left(T_i(\omega)x - h_i(\omega) \right)^- \right] \le \beta \right\}, \quad \beta \ge 0.$$

and

$$\text{L}_3(\lambda) \quad \min_{x\in X} \left\{ c(x) + Q_\lambda(x) \right\}, \quad \lambda \ge 0,$$

where

$$Q_\lambda(x) := \mathbb{E}_\omega \left[\min_y \{ qy : Wy = h(\omega) - T(\omega)x,\ y \ge 0 \} \right],$$

$$\begin{pmatrix} q \\ W \end{pmatrix} := \begin{pmatrix} \lambda & 0 \\ e & -I \end{pmatrix}, \quad e \in \mathbb{R}^m \text{ with } e_i = 1\ \forall i,$$

one easily verifies that $Q_\lambda(x) = \lambda \cdot \mathbb{E}_\omega \left[\max_i \left(T_i(\omega)x - h_i(\omega) \right)^- \right]$, so that Theorem 6.5.1 holds also for $\text{ICC}_3(\beta)$ and $\text{L}_3(\lambda)$.

6.5.1 Conclusions

In many cases, ICCs may be an appropriate tool for modeling random linear constraints. ICCs are more appropriate than CCs if the underlying concept 'risk := mean shortage' is more appropriate than 'risk := probability of positive shortage'. ICCs give rise to convex optimization problems, for all distributions of the random coefficients involved, matrix as well as right-hand side. Moreover, the feasibility sets defined by ICCs change continuously and strictly for all types of distributions of the random coefficients. Models with ICCs are more appropriate than recourse models, if one is not able to specify the penalty costs for infeasibilities. Unlike CC models, ICC models exhibit a natural mathematical equivalence with recourse models, provided by Lagrange multipliers. Computation of mean shortages is not easy in general; however, in simple situations where deterministic equivalent formulations for CCs exist, the same is true for ICCs, and as far as Monte Carlo simulation is concerned, there does not seem to be much difference in difficulty either. On the other hand, one might expect that progress in computation in simple and fixed recourse models can be used for computations with ICCs. Similarly, however, one might expect that progress in computation in mixed-integer linear programming leads to progress in computations with CCs.

6.6 Algorithms for Models with Integrated Chance Constraints

As discussed in the previous parts of this chapter, models with integrated chance constraints are convex in general, and in many cases a reduced form is known for the corresponding feasibility sets. In principle these models can therefore be solved by any standard optimization algorithm for (non-linear) convex problems.

However, a special purpose algorithm may take advantage from the specific structure of the problem at hand. Below we discuss such a special purpose algorithm for problems with an individual integrated chance constraint

$$\mathbb{E}_\omega\left[\eta(x,\omega)^-\right] \le \beta, \qquad \beta \in \mathbb{R}_+,$$

for the case that ω follows a finite discrete distribution, with $\Pr\{\omega = \omega^s\} = p^s$, $s \in S$. As before, we write $(T^s, h^s) = (T(\omega^s), h(\omega^s))$ for $s \in S$.

First we consider the following straightforward approach.

Exercise 6.6.1 Show that the following system of linear constraints is equivalent to an individual ICC (first type) with finite discrete distribution:

$$\begin{aligned} &T^s x + y^s \ge h^s, \; s \in S \\ &\sum_{s \in S} p^s y^s \le \beta \\ &y^s \ge 0, \qquad\qquad s \in S. \end{aligned} \tag{6.21}$$

◁

Hence, such an individual ICC can be represented in LP form at the cost of $|S|+1$ constraints and $|S|$ additional variables. Given the availability of very powerful LP software, the resulting LP problem can be solved efficiently if $|S|$ is not too large. Indeed, for small problem instances, this approach outperforms the special purpose algorithm that will be presented next. However, already for instances of moderate size, the algorithm below is (much) more efficient.

Exercise 6.6.2 Assume that ω follows a finite discrete distribution.

(i) Using (6.21), show that such an individual ICC is a relaxation of the corresponding individual chance constraint (for a suitable choice of β).
(ii) Prove an analogous result for the relation between a joint ICC and the corresponding joint CC.

◁

A more sophisticated algorithm is based on the reduced form of the induced feasible set, which according to Theorem 6.2.11 is given by

$$C_1(\beta) = \bigcap_{K \subset S} \left\{ x \in \mathbb{R}^n : \sum_{k \in K} p^k \left(h^k - T^k x\right) \le \beta \right\}.$$

This polyhedral convex set is defined by $2^{|S|} - 1$ linear constraints, which is already a huge number for small problem instances. Therefore, it is impractical or even impossible to solve the corresponding problem by a standard LP algorithm.

The main idea of Algorithm 6.6.4 below is that it uses only a partial description of the feasible set.

Remark 6.6.3 We assume that the problem

$$\begin{aligned} &\min_x \; cx \\ &\text{s.t.} \;\; x \in X \cap C^0, \end{aligned}$$

with $C^0 := \{x \in \mathbb{R}^n : \bar{T}x \geq \bar{h} - \beta\}$ (see Remark 6.2.12), is bounded. If not, this can be forced by including suitable simple bounds in the definition of the set X.

Algorithm 6.6.4 *Define the* current problem *CP at iteration t as*

$$\begin{aligned} &\min_x \; cx \\ &s.t. \;\; x \in X \\ &\qquad x \in C^t := \{x \in \mathbb{R}^n : d_i x \leq e_i, \quad i = 0, \ldots, t\} \end{aligned}$$

where the feasibility cuts $d_i x \leq e_i$, $i = 0, \ldots, t$, *give a partial description of the feasible set* $C_1(\beta)$. *Set* $t = 0$.
Iteration t:

(i) Solve the LP problem CP, giving an optimal solution x^t. *If CP is infeasible* STOP: *the problem is infeasible.*
(ii) Compute $\mathbb{E}_\omega\left[\eta(x^t, \omega)^-\right] = \sum_{s \in S} p^s \eta(x^t, \omega^s)^-$, *at the same time constructing the index set*

$$K^t := \left\{s \in S : \eta(x^t, \omega^s)^- > 0\right\}.$$

(iii) If $\mathbb{E}_\omega\left[\eta(x^t, \omega)^-\right] \leq \beta$ STOP: x^t *is an optimal solution. Otherwise, construct a feasibility cut* $d_{t+1}x \leq e_{t+1}$, *with*

$$d_{t+1} = -\sum_{k \in K^t} p^k T^k, \qquad e_{t+1} = \beta - \sum_{k \in K^t} p^k h^k,$$

and set

$$C^{t+1} = C^t \cap \left\{x \in \mathbb{R}^n : d_{t+1}x \leq e_{t+1}\right\}.$$

Proceed with iteration $t + 1$.

Theorem 6.6.5 *Consider the bounded optimization problem with individual ICC (first type)*

$$\begin{aligned} \min_x \ & cx \\ \text{s.t.} \ & x \in X \\ & \mathbb{E}_\omega \left[\eta(x, \omega)^- \right] \leq \beta, \end{aligned}$$

with β a given risk parameter. Assume that ω follows a finite discrete distribution. Then Algorithm 6.6.4 solves this problem in finitely many iterations.

Proof In each iteration, the algorithm either finds that the problem is infeasible, adds a new feasibility cut, or returns an optimal solution. The algorithm is finite since there are only finitely many feasibility cuts by Theorem 6.2.11. □

Example 6.6.6 Consider the ICC problem specified by $c = (-1 \ \ -2)$, $X = [0, 100]^2$, and the individual ICC constraint

$$\mathbb{E}_\omega \left[\left(\omega_1 x_1 + \omega_2 x_2 - \omega_3) \right)^- \right] \leq \beta,$$

where the 1000 realizations of $\left(T(\omega), h(\omega)\right) = \left(\omega_1, \omega_2, \omega_3\right)$ are sampled from the uniform distribution on $[-0.5, 0.5]^2 \times [0, 1]$, and $\beta = 9$.

The optimal solution $x^\star = (31.71, 65.58)$ is found by Algorithm 6.6.4 after ten iterations, i.e., using only nine feasibility cuts (out of the $2^{1000} \approx 10^{301}$ cuts constituting the reduced form of $C_1(\beta)$ according to Theorem 6.2.11). ◇

Chapter 7
Assignments

A1 Multiperiod inventory control
R1 Probabilistic interpretations of integrals of distribution functions
R2 Complete and sufficiently expensive recourse
R3 Simple recourse, only RHS random
R4 Workforce level planning: an example of multiple simple recourse?
C1 Chance constraints and integrated chance constraints
C2 The expected shortage function for positive random variables
C3 Convexity for a simple chance constraint
C4 Relation individual and joint chance constraints
C5 Joint Chance Constraints and Discrete Distributions
C6 Individual Chance Constraints with One Random Variable
U1 Stochastic Dominance and Expected Utility

7.1 A1: Multiperiod Inventory Control

Consider the inventory system from [31] with one product, over periods $t = 1, \ldots, s$, where stock keeping is possible but no backlogging. The production in period t is $x_t \geq 0$ and the demand in period t is $\omega_t \geq 0$, $t = 1, \ldots, s$. We assume that the demand is random. Denote the inventory at the end of period t by y_t, and the shortage at the end of period t by z_t. Obviously, y_t and z_t are nonnegative, and they depend on $x_1, \ldots, x_t$ and $\omega_1, \ldots, \omega_t$. In fact, we have, with $y_0 = 0$, the recursion

$$\begin{aligned} y_t &= (\omega_t - x_t - y_{t-1})^-, \quad t = 1, \ldots, s, \\ z_t &= (\omega_t - x_t - y_{t-1})^+, \quad t = 1, \ldots, s. \end{aligned} \tag{7.1}$$

Obviously, $y_t \cdot z_t = 0$ for all t.

W. K. Klein Haneveld et al., *Stochastic Programming*, Graduate Texts in Operations Research, https://doi.org/10.1007/978-3-030-29219-5_7

(a) Show that

$$y_t = \max_{j=1,\ldots,t+1} \sum_{k=j}^{t} (x_k - \omega_k) \qquad t = 1, \ldots, s,$$

where the empty sum ($\sim j = t+1$) is defined as zero.
Moreover, show that for fixed values of $(\omega_1, \ldots, \omega_t)$, y_t is a convex function of $(x_1, \ldots, x_t)$. Calculate the partial derivatives $\dfrac{\partial y_t}{\partial x_i}$, $i = 1, \ldots, s$.

We are interested in the total expected holding and shortage costs

$$Q(x) := \mathbb{E}_\omega \left[\sum_{t=1}^{s} \left(q_t^- y_t + q_t^+ z_t \right) \right] \tag{7.2}$$

where q_t^- (q_t^+, respectively) denotes the unit holding (shortage, respectively) cost in period t. In particular, we want to derive a suitable representation of the partial derivatives of Q as in (5.10) of [31] on p. 121.

(b) Eliminate the variables z_t in (7.2) using $z_t - y_t = \omega_t - y_{t-1}$ and show that, with $q_{s+1}^+ = 0$,

$$Q(x) = \sum_{t=1}^{s} q_t^+ \left(\mathbb{E}_\omega [\omega_t] - x_t \right) + \sum_{t=1}^{s} \left(q_t^+ + q_t^- - q_{t+1}^+ \right) \mathbb{E}_\omega [y_t] .$$

(c) Show, by differentiating under the integral (cf. (5.2)–(5.6) in [31])

$$\frac{\partial \mathbb{E}_\omega [y_t]}{\partial x_i} = P_{it}(x) := \begin{cases} \Pr\{y_i > 0,\, y_{i+1} > 0, \ldots, y_t > 0\}, & i = 1, \ldots, t \\ 0, & i = t+1, \ldots, s. \end{cases}$$

Consequently,

$$\frac{\partial Q}{\partial x_i}(x) = -q_i^+ + \sum_{t=i}^{s} \left(q_t^+ + q_t^- - q_{t+1}^+ \right) P_{it}(x)$$

To estimate these probabilities, take a sample of N (e.g. $N = 2000$) realizations of $\xi = (\xi_1, \ldots, \xi_s)$, and calculate for each realization (and any fixed x in which one is interested), all $y_1, y_2, \ldots, y_s$ using the recursion (7.1). Let $N_{it}(x)$ be the number of realizations with $y_i > 0$, $y_{i+1} > 0, \ldots$ *and* $y_t > 0$. Then,

$$P_{it}(x) \approx \frac{N_{it}(x)}{N}.$$

7.2 R1: Probabilistic Interpretations of Integrals of Distribution Functions

Let ω be a random variable with distribution function F, $F(t) := P\{\omega \leq t\}$, $t \in \mathbb{R}$. Then, for $x \in \mathbb{R}$,

$$G(x) := \mathbb{E}_\omega\left[(\omega - x)^+\right] = \int_x^\infty \left(1 - F(t)\right)\, dt,$$

$$H(x) := \mathbb{E}_\omega\left[(\omega - x)^-\right] = \int_{-\infty}^x F(t)\, dt.$$

This means, that the integral of the left tail of F can be interpreted as the expected value of the shortage of ω with respect to x, whereas the integral of the right tail of (the complement of) F gives the expected value of the surplus of ω with respect to x. See Sect. 3.3.3. This exercise shows, that also other integrals of F and $1 - F$ have interpretations as expected values.

(i) Show that for $a \in \mathbb{R}$ and $b \in [a, \infty]$,

$$\mathbb{E}_\omega\left[\min\left\{(\omega - a)^+, b - a\right\}\right] = \int_a^b \left(1 - F(t)\right)\, dt.$$

Give a formula for $\mu^+ := \mathbb{E}_\omega\left[(\omega)^+\right]$.

(ii) Show that for $b \in \mathbb{R}$ and $a \in [-\infty, b]$,

$$\mathbb{E}_\omega\left[\min\left\{(\omega - b)^-, b - a\right\}\right] = \int_a^b F(t)\, dt.$$

Give a formula for $\mu^- := \mathbb{E}_\omega\left[(\omega)^-\right]$.

From now on we assume that $\mu := \mathbb{E}_\omega[\omega]$ is finite, so that $\lim_{x \to +\infty} G(x) = \lim_{x \to -\infty} H(x) = 0$. Then, the integrals of G and H over the tails have probabilistic interpretations too.

(iii) Show that for $x \in \mathbb{R}$

$$\int_{-\infty}^x H(t)\, dt = \frac{1}{2}\mathbb{E}_\omega\left[\left((\omega - x)^-\right)^2\right]$$

$$\int_x^\infty G(t)\, dt = \frac{1}{2}\mathbb{E}_\omega\left[\left((\omega - x)^+\right)^2\right]$$

Note that for $x = \mu$ one gets expressions for the semi-variances of ω.

The following exercise deals with the probabilistic interpretation of the difference between the expected surplus function $G(x)$ and its lower bound $(\mu - x)^+$, and the difference between the expected shortage function $H(x)$ and its lower bound $(\mu - x)^-$. It appears, that both differences are equal, for all $x \in \mathbb{R}$; they provide a measure for the dispersion of the distribution of ω.

(iv) Show that

$$G(x) - (\mu - x)^+ = H(x) - (\mu - x)^- \quad \text{for all } x \in \mathbb{R}.$$

Denote this function of x by $m(x)$.

(v) Show that

(a) $m(x) = \min\{G(x), H(x)\} \quad \forall x \in \mathbb{R}$

(b) $\max_{x \in \mathbb{R}} m(x) = m(\mu) = \frac{1}{2}\mathbb{E}_\omega[|\omega - \mu|]$

(c) $\int_{-\infty}^{\infty} m(x)\,dx = \frac{1}{2}\mathbb{E}_\omega\left[(\omega - \mu)^2\right]$

7.3 R2: Complete and Sufficiently Expensive Recourse

This exercise deals with issues that come up in Sect. 3.2.3.

A standard recourse model of SLP can be written as

$$\min_{x \in \mathbb{R}^n} \{cx + Q(x) : Ax = b, x \geq 0\}$$

where

$$Q(x) := \mathbb{E}_\omega\left[v(h(\omega) - T(\omega)x)\right], \quad x \in \mathbb{R}^n$$

with

$$v(z) := \min_{y \in Y}\{qy : Wy = z\}, \quad z \in \mathbb{R}^m.$$

It aims at satisfying the random constraints $T(\omega)x = h(\omega)$ by repairing afterwards possible deviations $h(\omega) - T(\omega)x$ by means of recourse actions y. The function value $v(z)$ represents the minimum recourse cost necessary to repair the vector $z \in \mathbb{R}^m$ of deviations. We assume here that $Y = \mathbb{R}^p_+$, so that the underlying recourse structure is characterized completely by the $m \times p$ *recourse matrix* W and the $1 \times p$ row vector q of *unit recourse costs*.

7.3.1 *Finiteness of the Function v*

The matrix W is called a *complete recourse matrix* (= crm) if $\forall z \in \mathbb{R}^m : \exists y \in \mathbb{R}^p$ with $y \geq 0$ and $Wy = z$ (i.e. if $v(z) < \infty \ \forall z \in \mathbb{R}^m$). Moreover, the recourse cost vector q is called *extremely low* (in relation to a crm W) if $\exists z \in \mathbb{R}^m$ with $v(z) = -\infty$.
Consequently, $v(z)$ is finite for all $z \in \mathbb{R}^m$ if W is a crm and q is not extremely low.
By using linear programming duality (λ is a $1 \times m$ row vector)

$$\inf_{y \geq 0}\{qy : Wy = z\} = \sup_{\lambda}\{\lambda z : \lambda W \leq q\},$$

which is true unless both problems do not have feasible solutions, it is possible to give dual characterizations for these useful properties of the recourse structure (q, W).

(a) Show that W is crm iff $\{\lambda : \lambda W \leq q\}$ is a bounded set in $\mathbb{R}^m$ (or, equivalently, if $\lambda W \leq 0$ implies $\lambda = 0$).
(b) Show that q is not extremely low iff $\exists \lambda \in \mathbb{R}^m$ with $q \geq \lambda W$. In particular, any $q \geq 0$ is not extremely low.
(c) Show that each of the following matrices is a crm (directly, or by applying the dual characterization given under (a)).

   (i) $W = (I \ -I)$ simple recourse
   (ii) $W = (e \ -I)$ $e \in \mathbb{R}^m$, all components being 1

   For each of these case, what are the conditions for q to be not extremely low?

7.3.2 *Sufficiently Expensive Recourse*

Suppose that in the recourse model, with W a crm, the random elements $(T(\omega), h(\omega))$ are replaced by their mean values $(\bar{T}, \bar{h})$. Then we get the deterministic LP-problem

$$\text{(LP}_1\text{)} \quad \begin{array}{llll} \min\limits_{x,y} & cx & + qy & \\ & Ax & & = b \\ & \bar{T}x & + Wy & = \bar{h} \\ & x \geq 0 & y \geq 0. & \end{array}$$

Obviously, this is a relaxation of the LP-problem

$$\text{(LP}_0\text{)} \quad \min_x\{cx : Ax = b, \ \bar{T}x = \bar{h}, \ x \geq 0\}.$$

(d) Show that the optimal value of LP_1 is smaller than or equal to that of LP_0.

Assume that the optimal value of LP_0 is finite and denote the vector of shadow prices corresponding to the constraints $\bar{T}x = \bar{h}$ by $\bar{\lambda}$. (Assuming that the dual of LP_0 has a unique optimal solution, $\bar{\lambda}$ is a part of it.) Then we call the recourse structure (q, W) *sufficiently expensive* if $q > \bar{\lambda}W$. This name tries to reflect the fact, that for such values for q it is not optimal to use recourse actions in the deterministic linear program LP_1 even if they are available.

(e) Show that in LP_1 the optimal value for the recourse vector y is equal to zero, if the recourse is sufficiently expensive.

7.3.3 *Convexity and Differentiability*

Assume that q and W are such that v is finite everywhere.

(f) Show that v is a convex function. Using this, show that Q is a convex function.
(g) Show that v is piecewisely linear. Try to characterize its gradient in a fixed part of its domain where v is linear. (Hint: use duality.)
Guess a formula for the gradient of Q, assuming that it exists.

7.4 R3: Simple Recourse, Only RHS Random

Consider the recourse model

$$\min_{x \in \mathbb{R}^n} \{cx + Q(x) : Ax = b,\ x \geq 0\},$$

where

$$Q(x) := \mathbb{E}_\omega \left[v(h(\omega) - T(\omega)x) \right], \quad x \in \mathbb{R}^n,$$

with

$$v(z) := \min_{y \in \mathbb{R}^p} \{qy : Wy = z,\ y \geq 0\}, \quad z \in \mathbb{R}^m.$$

The special case, where $p = 2m$ and $\begin{pmatrix} q \\ W \end{pmatrix} = \begin{pmatrix} q^+ & q^- \\ I & -I \end{pmatrix}$ with $q^+ + q^- > 0$ is called *simple recourse*, as discussed in Sect. 3.3.2. The assumption 'only RHS random' means that T is deterministic, and only the components of h are random. In the rest of the exercise, we conveniently assume that $h_i(\omega) = \omega_i$, $i = 1, \ldots, m$.

(a) Show that with simple recourse and only RHS random the recourse model can be reformulated as the separable convex programming problem

$$\min_{x \in \mathbb{R}^n,\ s \in \mathbb{R}^m} \{cx + \sum_{i=1}^{m} Q_i(s_i) : Ax = b,\ Tx = s,\ x \geq 0\},$$

where

$$Q_i(s_i) := q_i^+(\mu_i - s_i) + (q_i^+ + q_i^-) \int_{-\infty}^{s_i} F_i(t)\, dt$$

with $\mu_i := \mathbb{E}_\omega[\omega_i]$ and $F_i(t) := \Pr\{\omega_i \leq t\}$.

Assume further, that each ω_i has a discrete distribution restricted to the values $\omega_i^1, \omega_i^2, \ldots, \omega_i^{K_i}$ with $\omega_i^1 < \omega_i^2 < \cdots < \omega_i^{K_i}$. Define

$$F_i^k := \Pr\left\{\omega_i \leq \omega_i^k\right\} = \sum_{h=1}^{k} \Pr\left\{\omega_i = \omega_i^h\right\}, \quad k = 0, \ldots, K_i,$$

so that $0 = F_i^0 < F_i^1 < \cdots < F_i^{K_i} = 1$, and

$$d_i^k = -q_i^+ + (q_i^+ + q_i^-)F_i^k, \quad k = 0, \ldots, K_i,$$

so that $-q_i^+ = d_i^0 < d_i^1 < \cdots < d_i^{K_i} = q_i^-$.

(b) Show that $Q_i(s_i)$ is a piecewise linear convex function on $\mathbb{R}$, with slope d_i^k on the interval $s_i \in [\omega_i^k, \omega_i^{k+1}]$, $k = 0, \ldots, K_i$ (where $\omega_i^0 := -\infty$, $\omega_i^{K_i+1} := +\infty$).
Sketch the graph of $Q_i(s_i)$.

(c) Show that the simple recourse problem, with only RHS random with discrete distributions, can be represented as the following LP-problem:

$$\begin{array}{llll}
\min\limits_{x \geq 0, u} & \sum\limits_{j=1}^{n} c_j x_j + \sum\limits_{i=1}^{m} \sum\limits_{k=0}^{K_i} d_i^k u_i^k & & \\
\text{s.t.} & \sum\limits_{j=1}^{n} a_{lj} x_j & = b_l, & l = 1, \ldots, m_1 \\
& \sum\limits_{j=1}^{n} t_{ij} x_j - \sum\limits_{k=0}^{K_i} u_i^k & = \mu_i, & i = 1, \ldots, m \\
& u_i^0 \leq \omega_i^1 - \mu_i, & & i = 1, \ldots, m \\
& 0 \leq u_i^k \leq \omega_i^{k+1} - \omega_i^k, & & i = 1, \ldots, m,\ k = 1, \ldots, K_i - 1 \\
& 0 \leq u_i^{K_i} & & i = 1, \ldots, m
\end{array}$$

Hint: Show that the part of the model, determined by the variables u_i^k, $k = 1, \dots, K_i$ precisely gives the contribution

$$Q_i(\mu_i + \sum_{k=0}^{K_i} u_i^k)$$

to the objective function. In order to see this, note that since $d_i^0 < d_i^1 < \cdots < d_i^{K_i}$, a variable u_i^k will only be positive in the optimum if its predecessor u_i^{k-1} is at its maximum value!

7.5 R4: Workforce Level Planning: An Example of Multiple Simple Recourse?

Let ω be a nonnegative variable with cumulative distribution function F, that is strictly increasing and continuous. Let c, q_1, q_2 be real numbers such that $0 < c < q_1 < q_2$, and let γ be a fixed number in $(0, 1)$. Then we consider the following recourse problem:

$$\text{(RP)} \qquad \min_{x \in \mathbb{R}} \{C(x) : x \geq 0\}$$

where $C(x) := cx + Q(x)$ with $Q(x) := \mathbb{E}_\omega [g(\omega, x)]$, and

$$\begin{aligned} g(\omega, x) := \min_{y \in \mathbb{R}^2} \; & q_1 y_1 + q_2 y_2 \\ \text{s.t.} \; & y_1 + y_2 \geq \omega - x \\ & y_1 \leq \gamma \cdot x \\ & y_1 \geq 0 \quad y_2 \geq 0 \end{aligned}$$

Interpretation This model is a prototype for the following situation. At the beginning of a year, a company has to decide on the level of the regular workforce. At that stage, the amount of work is not yet known precisely. If the actual work happens to be more than can be handled by the regular staff, then there are two recourse actions. First, there is a restricted possibility for overwork. Secondly, ad hoc workforce can be hired from an outside agency. Of course, overwork has to be paid for, whereas hired workforce is still more expensive.

(i) Indicate why the recourse structure in this model is relatively complete, but not complete. Is it 'multiple simple'?
(ii) Solve the second-stage problem analytically for all values of ω and x, and draw a graph of the function $g(\omega, \cdot)$ as a function of $x \in [0, \infty)$ for a fixed positive value of ω.

(iii) Show that

$$Q(x) = q_1\mathbb{E}_\omega\left[(\omega - x)^+\right] + (q_2 - q_1)\mathbb{E}_\omega\left[(\omega - (1+\gamma)x)^+\right].$$

Why is this function convex and differentiable?

(iv) Calculate $C'(x)$, and characterize the optimal solution x^* of RP.

(v) Show that $(1+\gamma)^{-1}x_0 \le x^* \le x_0$, where

$$x_0 := F^{-1}\left(1 - \frac{c}{q_2 + (q_2 - q_1)\gamma}\right).$$

Hint: Show that

$$C'(x) \ge c - \big(q_2 + (q_2 - q_1)\gamma\big) \cdot \big(1 - F(x)\big),$$

hence $C'(x_0) \ge 0$; and

$$C'(x) \le c - \big(q_2 + (q_2 - q_1)\gamma\big) \cdot \big(1 - F(x(1+\gamma))\big),$$

hence $C'((1+\gamma)^{-1}x_0) \le 0$.

(vi) Define

$$\hat{x} := \frac{1+\gamma/2}{1+\gamma}x_0.$$

Give an upper bound on the absolute value of the relative error, when $\hat{x}$ is used as an approximation for x^*.

7.6 C1: Chance Constraints and Integrated Chance Constraints

In Sect. 5.2.2 various examples of CCs have been discussed. Here, another example is presented, with emphasis on the differences with ICCs, as was the case in Sect. 6.2.2.

The goal to choose $x = (x_1, \ldots, x_n) \in \mathbb{R}^n$ such that $\sum_{j=1}^n \omega_j x_j \ge \omega_{n+1}$ in case $\omega_1, \ldots, \omega_{n+1}$ are random variables, is often formulated in terms of a chance constraint. The corresponding set of feasible points is then

$$C_0(\alpha) := \left\{x \in \mathbb{R}^n : \Pr\left\{\sum_{j=1}^n \omega_j x_j \ge \omega_{n+1}\right\} \ge \alpha\right\}, \quad \alpha \in [0, 1],$$

where $1-\alpha$ denotes the maximal acceptable *probability of shortage*. An alternative specification of the set of feasible points is

$$C_1(\beta) := \left\{ x \in \mathbb{R}^n : \mathbb{E}_{a,b}\left[\left(\sum_{j=1}^n \omega_j x_j - \omega_{n+1} \right)^- \right] \le \beta \right\}, \qquad \beta \in [0, \infty],$$

with β the maximal *expected shortage*. The examples below are intended to illustrate that $C_1(\beta)$ has 'nicer' mathematical properties than $C_0(\alpha)$:

- $C_0(\alpha)$ is not necessarily convex, $C_1(\beta)$ is.
- $C_1(\beta)$ increases smoothly with increasing β, whereas $C_0(\alpha)$ may increase stepwise.

(i) Calculate $C_0(\alpha)$ en $C_1(\beta)$ for all values of α, β for the instance with $n = 2$, $\Pr\{(\omega_1, \omega_2, \omega_3) = (-2, 1, 0)\} = \Pr\{(\omega_1, \omega_2, \omega_3) = (1, -2, 0)\} = \frac{1}{2}$. Do the same for
$n = 2$, $\Pr\{(\omega_1, \omega_2, \omega_3) = (0, 3, 6)\} = \Pr\{(\omega_1, \omega_2, \omega_3) = (3, 0, 6)\} = 0.1$, and
$\Pr\{(\omega_1, \omega_2, \omega_3) = (2, 2, 6)\} = 0.8$. Draw pictures too.

(ii) Prove that for every distribution of $(\omega_1, \ldots, \omega_{n+1})$ and every $\beta \ge 0$ the set $C_1(\beta)$ is convex.

7.7 C2: The Expected Shortage Function for Positive Random Variables

Let ω be a nonnegative random variable, discretely or continuously distributed. Define $F(t) := P\{\omega < t\}$, $t \in \mathbb{R}$, $\mu := \mathbb{E}_\omega[\omega] \in (0, \infty)$, $\sigma^2 := \text{var}(\omega)$. Let $G : \mathbb{R} \mapsto \mathbb{R}$ be the function defined by

$$G(x) = \begin{cases} 0 & , x < 0, \\ \frac{1}{\mu}\mathbb{E}_\omega[\min(x, \omega)] & , x \ge 0. \end{cases}$$

(i) Show that G is a cumulative distribution function of a nonnegative random variable.

(ii) Show that G has the density g, given by

$$g(x) = \begin{cases} \dfrac{1}{\mu}(1 - F(x)), & x \ge 0 \\ 0, & x < 0. \end{cases}$$

(iii) Show that the mean value of G is given by $\dfrac{\mu^2 + \sigma^2}{2\mu}$.

(iv) Show that $G = F$ if F is the distribution function of an exponential distribution.

(v) Show, that the integrated chance constraint (see (6.5) in Sect. 6.2)

$$\mathbb{E}_\omega\left[(\omega - x)^+\right] \le \alpha \mathbb{E}_\omega\left[\omega\right], \quad \alpha \in (0, 1)$$

is equivalent to $x \ge G^{-1}(1-\alpha)$.

7.8 C3: Convexity for a Simple Chance Constraint

Let ω_1 and ω_2 be independent normal random variables, with mean values μ and ν, and variances σ^2 and τ^2, respectively. Define

$$C(\alpha) := \{x \in \mathbb{R} : \Pr\{\omega_1 x \ge \omega_2\} \ge \alpha\}, \quad 0 \le \alpha \le 1.$$

Show that $C(\alpha)$ is *not* convex if and only if the following conditions are satisfied:

$$\sigma > 0,\ \nu > 0 \text{ and } -\sqrt{\frac{\mu^2}{\sigma^2} + \frac{\nu^2}{\tau^2}} < \Phi^{-1}(\alpha) < \frac{-\mid \mu \mid}{\sigma},$$

where Φ is the cdf of the standard normal distribution and $\nu^2/\tau^2 := \infty$ if $\tau = 0$. What does $C(\alpha)$ look like under these conditions?
An extreme case is: $\mu = \tau = 0$, $\sigma = \nu = 1$.

Remarks

(1) The theory (Theorem 5.2.8 and Corollary 5.2.21) indicates that under general conditions $C(\alpha)$ is a closed set, and convex for $\alpha \ge \frac{1}{2}$ (but possibly empty).

(2) In principle there are many specific cases to consider, not all of them very interesting: $\alpha = 0$, $0 < \alpha < 1$, $\alpha = 1$; $\mu < 0$, $\mu = 0$, $\mu > 0$; $\nu < 0$, $\nu = 0$, $\nu > 0$; $\sigma = 0$, $\sigma > 0$; $\tau = 0$, $\tau > 0$.

7.9 C4: Relation Individual and Joint Chance Constraints

Let $T = T(\omega) = (t_{ij}(\omega))$ be a stochastic $m \times n$ matrix and $h = h(\omega) = (h_i(\omega))$ a stochastic $m \times 1$ vector. As in Sect. 5.1, for every $x \in \mathbb{R}^n$ we define the individual reliabilities

$$p_i(x) := \Pr\left\{\sum_{j=1}^{n} t_{ij}(\omega) x_j \ge h_i(\omega)\right\}, \quad i = 1, \ldots, m,$$

and the joint reliability

$$p(x) := \Pr\{T(\omega)x \geq h(\omega)\}.$$

Individual chance constraints lead to

$$p_i(x) \geq \alpha_i, \quad i = 1, \ldots, m,$$

and a joint chance constraint is

$$p(x) \geq \alpha.$$

The associated feasibility sets are

$$C_i(\alpha_i) := \{x \in \mathbb{R}^n : p_i(x) \geq \alpha_i\}, \alpha_i \in (0,1), i = 1, \ldots, m,$$

$$C(\alpha) := \{x \in \mathbb{R}^n : p(x) \geq \alpha\}, \quad \alpha \in (0,1).$$

(i) Under which condition(s) holds $p(x) = \prod_{i=1}^m p_i(x)$? Verify, that in that case

$$\bigcap_{i=1}^m C_i(\alpha_i) \subset C(\prod_{i=1}^m \alpha_i).$$

Why does equality not hold here in general?

(ii) Show that in general:

$$\bigcap_{i=1}^m C_i\left(1 - \frac{1-\alpha}{m}\right) \subset C(\alpha) \subset \bigcap_{i=1}^m C_i(\alpha), \quad \alpha \in (0,1).$$

Hint: First prove the *Bonferroni inequality*: for arbitrary events $E_1, \ldots, E_m$ it holds $P(\bigcap_{i=1}^m E_i) \geq 1 - \sum_{i=1}^m (1 - P(E_i))$.

(iii) The first inclusion under (ii) allows to find a point satisfying the joint chance constraint, by only considering individual chance constraints. Show this for the special case:

$$T \text{ deterministic}, \quad h \sim N_m(\mu, V).$$

Hint: Show that for values of $\tau \in \mathbb{R}_+$ to be specified,

$$\{x \in \mathbb{R}^n : Tx \geq \mu + \tau s\} \subset C(\alpha) \text{ where } s \in \mathbb{R}^m, \; s_i := (V_{ii})^{\frac{1}{2}}.$$

7.10 C5: Joint Chance Constraints, Only RHS Random, Discrete Distribution

Consider SLPwJCC where $p(x) = \Pr\{Tx \geq \omega\}$ and ω is a random vector in $\mathbb{R}^m$, $m \geq 2$, with finite support $\Omega = \{\omega^1, \ldots, \omega^K\}$, $\Pr\left\{\omega = \omega^k\right\} = p_k$, $k = 1, \ldots, K$, and cdf F. (See Sect. 5.3 for the more general case '$T(\omega)$ is random').

Since T is fixed, we may define the tender variables $s := Tx$ and the reliability function of s as $\bar{p}(s) := \Pr\{s \geq \omega\}$. The goal of this exercise is to show that the corresponding feasible sets

$$\bar{C}(\alpha) := \left\{s \in \mathbb{R}^m : \bar{p}(s) \geq \alpha\right\}, \qquad 0 < \alpha \leq 1,$$

can be represented as unions of translated positive orthants in $\mathbb{R}^m$:

$$\begin{aligned} &\text{For all } \alpha \in (0,1)\ \exists \text{ a finite set } H_\alpha \subset \mathbb{R}^m,\ H_\alpha = \left\{h^j, j \in J(\alpha)\right\}, \\ &\text{such that} \\ &\qquad \bar{p}(s) \geq \alpha \Longleftrightarrow \exists j \in J(\alpha) : s \geq h^j \end{aligned} \tag{7.3}$$

That is,

$$\bar{C}(\alpha) = \bigcup_{j \in J(\alpha)} \left\{h^j + \mathbb{R}^m_+\right\}.$$

(i) *Example* $m = 2$, $K = 4$, $\omega^1 = (1, 4)$, $\omega^2 = (2, 2)$, $\omega^3 = (3, 3)$, $\omega^4 = (4, 1)$, $p_1 = p_2 = p_3 = p_4 = 1/4$.
By making pictures, determine for this example the sets H_α, $\alpha \in (0, 1]$.
Is it true that $H_\alpha \subset \Omega$?

(ii) **Definition 7.10.1** For $0 < \alpha \leq 1$, a point $h \in \mathbb{R}^m$ is called an *α-efficient point of the distribution F of ω* if $F(h) \geq \alpha$ and there is no $z \leq h$, $z \neq h$, such that $F(z) \geq \alpha$.
Show that for all $\alpha \in (0, 1)$ the set of α-efficient points is non-empty and finite. (Remark: h can only be efficient if for each i, $i = 1, \ldots, m$, there is a $k_i \in \{1, \ldots, K\}$ such that $h_i = (\omega^{k_i})_i$.)

(iii) Show that (7.3) is true if one specifies $H_\alpha :=$ set of all α-efficient points of F.

Remark In practical problems it is not easy to calculate the relevant efficient set of F.

7.11 C6: Individual Chance Constraints with One Random Variable

In this exercise we study the set of feasible points determined by an individual chance constraint in which only one random variable occurs, by means of a few examples, as is done in Sect. 5.2.2. All deal with decision vectors $x = (x_1, x_2) \in \mathbb{R}^2$. In the first case only the RHS is random, whereas in the other cases at least one coefficient of a decision variable is random. Let ω be a random variable, with cumulative distribution function $F : F(t) := \Pr\{\omega \leq t\}, t \in \mathbb{R}$. For simplicity, we assume that F is continuous and strictly increasing on a bounded interval $[r, s]$, with $F(r) = 0$, $F(s) = 1$. If needed, one may assume $r = 0$ and $s = 3$ and that ω is uniformly distributed. Let $a_1(\omega), a_2(\omega), b(\omega)$ be coefficients that depend linearly on ω:

$$\left(a_1(\omega), a_2(\omega), b(\omega)\right) = \left(a_1^0, a_2^0, b^0\right) + \omega\left(a_1^1, a_2^1, b^1\right)$$

and define the reliability $p(x)$ of any x in $\mathbb{R}^2$ by

$$p(x) := \Pr\{a_1(\omega)x_1 + a_2(\omega)x_2 \geq b(\omega)\}.$$

Then the corresponding sets of feasible solutions are

$$C(\alpha) := \{x \in \mathbb{R}^2 : p(x) \geq \alpha\}, \quad \alpha \in [0, 1].$$

Case 1 $(a_1(\omega), a_2(\omega), b(\omega)) = (1, 2, \omega)$ so that $p(x) = \Pr\{x_1 + 2x_2 \geq \omega\}$.

(i) Express $p(x)$ in terms of F. Show that p is a continuous function.
(ii) Find an explicit representation of $C(\alpha)$ in terms of F^{-1}, for all $\alpha \in [0, 1]$. Show that $C(\alpha)$ is a polyhedral set, hence convex and closed. Make pictures of the extremely risky set $\{x \in \mathbb{R}^2 : p(x) = 0\}$, and of $C(0)$, $C(\frac{1}{4})$, $C(\frac{1}{2})$, $C(\frac{3}{4})$ and $C(1)$, the last being the extremely safe set.

Case 2 $(a_1(\omega), a_2(\omega), a_3(\omega)) = (\omega, 1, 2)$ so that $p(x) = \Pr\{\omega x_1 + x_2 \geq 2\}$.

(i) Express $p(x)$ in terms of F. (Consider three cases: $x_1 < 0, x_1 = 0, x_1 > 0$). Show that p is continuous at each $x \in \mathbb{R}^2$, except at $(x_1, x_2) = (0, 2)$. Verify that p is upper semi-continuous at this point.
(ii) Find an explicit representation of $C(\alpha)$ in terms of F^{-1}, for all $\alpha \in [0, 1]$. Show that $C(\alpha)$ is convex if $\alpha = 0$ and if $\alpha \in [\frac{1}{2}, 1]$, but nonconvex for $\alpha \in (0, \frac{1}{2})$. Make pictures of the extremely risky set $\{x \in \mathbb{R}^2 : p(x) = 0\}$ and of $C(0)$, $C(\frac{1}{4})$, $C(\frac{1}{2})$, $C(\frac{3}{4})$ and $C(1)$, the last being the extremely safe set.

Case 3 $(a_1(\omega), a_2(\omega), b(\omega)) = (\omega, -1, 3\omega - 2)$ so that $p(x) = \Pr\{\omega x_1 - x_2 \geq 3\omega - 2\}$.

(i) Express $p(x)$ in terms of F (consider the cases $x_1 < 3, x_1 = 3, x_1 > 3$ separately). Show that $p(x)$ is continuous at each $x \in \mathbb{R}^2$, except at $(x_1, x_2) = (3, 2)$. Verify that p is upper semi-continuous at this point.
(ii) Answer the same question as under Case 2, (ii).

7.12 U1: Stochastic Dominance and Expected Utility

This exercise relates the theory of Expected Utility in Chap. 2 with the concept of Stochastic Dominance.

For $I := [a, b]$ a bounded interval in $\mathbb{R}$, denote the family of all cumulative distribution functions of probability distributions with support in I by $\mathcal{F}$. A random variable Z with a distribution function in $\mathcal{F}$ will be called a random *yield*. In $\mathcal{F}$, we define two partial orderings, as follows.

- For F_1 and F_2 in $\mathcal{F}$, we say that F_1 *dominates* F_2 *stochastically in the first degree* (Notation $F_1 \overset{1}{\succ} F_2$) if

$$F_1(t) \le F_2(t) \text{ for all } t \in \mathbb{R},$$
$$F_1(t) < F_2(t) \text{ for some } t \in \mathbb{R}.$$

- For F_1 and F_2 in $\mathcal{F}$, we say that F_1 *dominates* F_2 *stochastically in the second degree* (Notation $F_1 \overset{2}{\succ} F_2$) if

$$IF_1(t) \le IF_2(t) \text{ for all } t \in \mathbb{R},$$
$$IF_1(t) < IF_2(t) \text{ for some } t \in \mathbb{R},$$

where $IF(t) := \int_{-\infty}^{t} F(x)dx,\ t \in \mathbb{R}$. Hence $IF(t) = H(t) = \mathbb{E}_Z\left[(Z-t)^-\right]$, see Sect. 3.3.2.

In this exercise it is asked to show, that large groups of expected utility maximizers (see Chap. 2) agree that stochastic dominant yield distributions are preferred.

(a) Suppose that $F_1 \in \mathcal{F}$, $F_2 \in \mathcal{F}$ and $F_1 \overset{1}{\succ} F_2$

(i) Show that $\mu_1 > \mu_2$, where μ_i is the mean value of F_i. That is, the risk-neutral decision maker prefers F_1 to F_2.
Hint: use $\mu_i - a = \int_a^b (1 - F_i(t))dt$ (to be shown by partial integration, or by using the results on $G(x) = \mathbb{E}_Z\left[(Z-x)^+\right]$ in Sect. 3.3.2):

$$\mu_i - a = \mathbb{E}_{F_i}\left[(Z-a)\right] = \mathbb{E}_{F_i}\left[(Z-a)^+\right]$$
$$= \int_a^{\infty} \left(1 - F_i(t)\right) dt = \int_a^b \left(1 - F_i(t)\right) dt.$$

(ii) Show, that no reliability maximizer prefers F_2 to F_1. Here, a reliability maximizer is an expected utility maximizer, using the function

$$u_x(z) = \begin{cases} 0, \text{ if } z < x \\ 1, \text{ if } z \geq x \end{cases}$$

as utility function, for some fixed $x \in \mathbb{R}$.

(iii) Let $\mathcal{U}_1$ be the class of all strictly increasing differentiable real functions on I.

1. Show that for all $u \in \mathcal{U}_1$ and $F \in \mathcal{F}$

$$\mathbb{E}_F\left[u(Z)\right] = u(a) + \int_a^b \big(1 - F(t)\big)\, u'(t)\, dt.$$

Hint: use partial integration.

2. Show that all expected utility maximizers, using a utility function $u \in \mathcal{U}_1$, agree that F_1 is strictly preferred to F_2.
NB: Risk averters and risk seekers agree about this!

(b) Suppose that $F_1 \in \mathcal{F}$, $F_2 \in \mathcal{F}$ and $F_1 \overset{2}{\succ} F_2$.

(i) Show that second degree stochastic dominance is implied by first degree stochastic dominance, but not conversely.

(ii) Show that no expected shortage minimizer prefers F_2 to F_1. Here, an expected shortage minimizer is an expected utility maximizer, using the function

$$u_x(z) = -(z - x)^-, \quad z \in \mathbb{R},$$

as utility function, for some fixed $x \in \mathbb{R}$.
Hint: use $IF(x) = \mathbb{E}_F\left[(Z - x)^-\right]$, $x \in \mathbb{R}$.

(iii) Let $\mathcal{U}_2$ be the class of all strictly increasing *twice* differentiable real functions on I.

1. Show that for all $u \in \mathcal{U}_2$ and $F \in \mathcal{F}$

$$\mathbb{E}_F\left[u(Z)\right] = u(b) - u'(b)IF(b) + \int_a^b IF(t)u''(t)\, dt.$$

Hint: First show by partial integration

$$\mathbb{E}_F\left[u(Z)\right] = u(b) - \int_a^b F(t)u'(t)\, dt.$$

Then apply partial integration again, using $(IF)'(x) = F(x)$.

2. Show that all expected utility maximizers, using a *concave* utility function $u \in \mathcal{U}_2$, agree that F_1 is strictly preferred to F_2.
NB: All risk averters agree about this!

Chapter 8
Case Studies

8.1 Multi-Period Production Planning[1]

Consider a factory that produces several different products. To produce these products, machines and labor (skilled and unskilled) are needed. It is assumed that stock keeping is possible, but backlogging is not. For every product it is known how many machine-hours and hours of labor are needed. At the end of each period, the demands, which are not known with certainty, must be met. Costs are induced when inventory is either too large or too small to satisfy demand, namely inventory and purchase costs respectively. To satisfy demand in a period, additional labor or machine hours can be used, although these additions are bounded. Costs are associated with increasing or decreasing the work force level. (See Assignment A1 for a simplified version of this problem.)

In the following we will look at a recourse model representing this problem. The parameters and variables are described in the list on the next page.

In our recourse model the first-stage decisions will deal with the number of each product to be produced in each period, the extra capacity to be used of each source in every period and the increases and decreases of the work force level in each period. The second-stage will be concerned with the number of products to be held in store or the number of products to be bought elsewhere to meet demand. This leads to the following second-stage constraints:

- $x_{it} + y^-_{i,t-1} + y^+_{it} - y^-_{it} = \xi_{it}, \quad i = 1, \ldots, n, \;\; t = 1, \ldots, s$

  In each period, for each product, demand has to be met, by a combination of production and purchases in that period and inventories from previous periods. Surpluses have to be stored.

[1]See Peters et al. [31].

W. K. Klein Haneveld et al., *Stochastic Programming*, Graduate Texts in Operations Research, https://doi.org/10.1007/978-3-030-29219-5_8

Indices:

t	Period
i	Product
j	Source of production

Numbers:

s	The number of periods considered
n	The number of different products produced by the factory
m	The number of different sources of production

First-stage decision variables:

x_{it}	Production of product i in period $t, i = 1, \ldots, n, t = 1, \ldots, s$
u_{jt}	Additional capacity of source j in period t, $j = 1, \ldots, m, t = 1, \ldots, s$
$z^+_{t-1,t}, z^-_{t-1,t}$	Planned increase/decrease of work force level between periods t and $t-1, t = 2, \ldots, s$

Second-stage decision variables:

y^-_{it}	*Surplus* of product i at the end of period t, $i = 1, \ldots, n, t = 1, \ldots, s$
y^+_{it}	*Shortage* of product i at the end of period t, $i = 1, \ldots, n, t = 1, \ldots, s$

Random variables:

ξ_{it}	Random demand for product i in period $t, i = 1, \ldots, n, t = 1, \ldots, s$

Deterministic parameters:

c^1_{it}	Unit cost of producing product i in period $t, i = 1, \ldots, n, t = 1, \ldots, s$
c^2_{jt}	Unit cost of extra capacity of source j in period t, $j = 1, \ldots, m, t = 1, \ldots, s$
$c^{3+}_{t-1,t}, c^{3-}_{t-1,t}$	Unit cost of increasing/decreasing work force level between periods t and $t-1, t = 2, \ldots, s$
q^+_{it}, q^-_{it}	Unit cost of shortages/surpluses of production of product i in period $t, i = 1, \ldots, n, t = 1, \ldots, s$
up_{jt}	Upper bounds for the use of extra capacity of source j in period t, $j = 1, \ldots, m, t = 1, \ldots, s$
a_{ij}	Number of units of source j needed for the production of one Unit of product $i, i = 1, \ldots, n, j = 1, \ldots, m$
b_{jt}	Number of units of source j available in period t, $j = 1, \ldots, m, t = 1, \ldots, s$

The first-stage constraints are:

- $\sum_{i=1}^{n} a_{ij} \cdot x_{it} \leq b_{jt} + u_{jt}, \quad j = 1, \ldots, m, \quad t = 1, \ldots, s$

In each period the amount of available sources must be sufficient to produce the desired number of products.

- $z^{+}_{t-1,t} - z^{-}_{t-1,t} = \sum_{i=1}^{n} a_{im} \cdot (x_{it} - x_{i,t-1}), \quad t = 2, \ldots, s$
  The definition of the planned in/decrease of the work force level, assuming the m-th production source is labor.
- $u_{jt} \leq up_{jt}, \quad j = 1, \ldots, m, \; t = 1, \ldots, s$
  The upper bounds on the extra capacity of the sources.

The recourse problem looks as follows:

$$\min_{x,u,z} \left\{ \sum_{i=1}^{n} \sum_{t=1}^{s} c^{1}_{it} x_{it} + \sum_{j=1}^{m} \sum_{t=1}^{s} c^{2}_{jt} u_{jt} + \sum_{t=2}^{s} c^{3+}_{t-1,t} z^{+}_{t-1,t} + \sum_{t=2}^{s} c^{3-}_{t-1,t} z^{-}_{t-1,t} + \right.$$

$$\left. \mathbb{E}_{\xi} \left[\min_{y} \left(\sum_{i=1}^{n} \sum_{t=1}^{s} q^{+}_{it} y^{+}_{it} + \sum_{i=1}^{n} \sum_{t=1}^{s} q^{-}_{it} y^{-}_{it} \right) \right] \right\}$$

subject to the restrictions stated above and non-negativity of all variables.

We will consider two products and three sources of production, of which the last one deals with labor. We will look at a four-period planning horizon. The initial inventories are taken to be zero.

The data are described in Tables 8.1, 8.2, 8.3 and 8.4. Table 8.1 gives the means and *standard deviations* of the normal distributions of the demand in each period. Table 8.2 shows the capacities of the sources needed to produce one unit of each product. Table 8.3 describes the capacities of the sources that are available in each period, and also the upper bounds of the possible extra capacity. Table 8.4 describes the unit costs associated with the various actions.

(a) Compute (i) the Expected Value Solution, and (ii) the Expected Result (of the expected value solution). Give an interpretation of the outcomes. In particular, discuss the usefulness of solving the recourse model.
Solve the recourse model and compare the outcome with (i) and (ii) above. In particular, discuss the so-called Value of the Stochastic Solution.

Table 8.1 Means and standard deviations of the distributions of demand

Period	Product 1	Product 2
1	(300, 45)	(500, 75)
2	(320, 45)	(500, 75)
3	(440, 45)	(500, 75)
4	(480, 45)	(600, 75)

Table 8.2 Capacity of sources needed for the products

Source	Product 1	Product 2
Machine 1	4	5
Machine 2	3	2
Labor	3	7

Table 8.3 Capacity of source available, and maximum possible level of increase

Period	Machine 1	Machine 2	Labor
1	4000 (400)	3000 (300)	4500 (450)
2	4000 (400)	3000 (300)	4500 (450)
3	4000 (400)	2500 (250)	3750 (375)
4	3500 (350)	3000 (350)	3500 (350)

Table 8.4 Cost coefficients

Source of cost	Unit cost	Source of cost	Unit cost
Add. capacity of machine 1	15	Inventory holding, period 1, 2, 3	
Add. capacity of machine 2	20	Product 1	25
Add. capacity of labor	10	Product 2	30
Production of product 1	100	Inventory holding, period 4	
Production of product 2	150	Product 1	100
Increase work force level	20	Product 2	150
Decrease work force level	15	Purchases from outside	
		Product 1	400
		Product 2	450

Compute the Wait-and-See solution and compare the outcome with previous results. In particular, discuss the so-called Expected Value of Perfect Information.

(b) Analyze the sensitivity of the problem with respect to the upper bounds on the expansion of the capacity of labor and the unit cost of additional capacity of labor.
You could also look at the sensitivity to the assumptions on the demand for the products (independence of demand for the two products; independence over time).

(c) For the optimal solution of the recourse model, found under (a), calculate the probability that purchases from outside are not needed in any period and for any product.

(d) We will now look at the problem as a chance-constrained problem. Our model will now be a joint chance-constrained model which specifies the reliability α that the production plan is sufficient to satisfy demand for both products in all periods.
Take into account the purchases from outside by assuming that these have to be ordered in advance, so they are specified as first-stage variables in this model. The cost associated with these purchases will be lower than in the recourse model, because you order in advance. Also the purchase cost may vary over time. The disadvantage is of course that you **have** to buy what you order.
As for storage, you will also have to build in this option, although you cannot assign cost to it. (So do not define first-stage variables to represent storage, find another way!)
Solve the model for different values of α. Compare the results with the expected value model and the recourse model.

(e) Discuss the model.

8.2 Water Management in the Bodrog River Area[2]

Consider a water resources system containing amongst others reservoir V. In principle, the capacity of V can be increased, at rather large costs. The question that is asked is whether the existing system can meet all the requirements. And if so, with which reliability? To what extent should the capacity of V be increased, if at all?

The main purposes of the water resources system are the irrigation water supply, industrial uses, flood control, environmental conservation and recreation. To decide whether the existing system can meet all the requirements, we will look at an aggregated stochastic model in which the monthly flow and irrigation water requirements are aggregated into four periods:

- The *first* period, November until April, comprises the winter and spring periods, filling the reservoir.
- The *second* period, May and June, includes irrigation and industrial demands, as does the *fourth* period, September and October.
- The *third* period, July and August, includes in addition the recreation demands.

Requirements for the minimum pool due to environmental control and enhancement and flood control pertain to all the periods.

In the following we will look at a chance-constrained model for the system. As our objective we should take the minimization of the cost of increasing the capacity of reservoir V. As the cost is an increasing function of the reservoir capacity, we will minimize the capacity of V instead.

We define the following variables (unit is Mm^3):

x_0	Required capacity of reservoir V
x_i	Total planned release in period i, $i = 1, \ldots, 4$
s_i	Reservoir storage at the end of period i, $i = 1, \ldots, 4$
β_i	Random demand (irrigation) in period i, $i = 2, 3, 4$
r_i	Random water inflow in period i, $i = 1, \ldots, 4$
ζ_i	Accumulated water inflow up to period i, $i = 1, \ldots, 4$; $\zeta_i = \sum_{j=1}^{i} r_j$
d_i	Deterministic demand (min. flow, industry) in period i, $i = 1, \ldots, 4$
s_0	Initial reservoir storage
m_i	Prescribed minimum storage for period i, $i = 1, \ldots, 4$
v_i	Prescribed minimum freeboard volume for period i, $i = 1, \ldots, 4$
$\alpha_i, \gamma_i, \alpha$	Probability levels representing minimum required reliability levels (for period i, $i = 1, \ldots, 4$)

[2]For further information about this case, see *Stochastic programming in water management: A case study and a comparison of solution techniques*, Dupačova et al., EJOR 52 (1991) 28–44.

The following constraints are formulated:

- $P(x_i \geq \beta_i + d_i,\ i = 2, 3, 4) \geq \alpha$
  A joint chance constraint bounding the chance that demand will exceed total release. In addition to this joint chance constraint a separate constraint for the first period (period without stochastic demand) will be added:
  $x_1 \geq d_1$
- $P(s_i \geq m_i) \geq \alpha_i, \quad i = 1, \ldots, 4$
  Individual chance constraints considering the minimum storage
- $P(s_i + v_i \leq x_0) \geq \gamma_i, \quad i = 1, \ldots, 4$
  Individual chance constraints considering the minimum freeboard volume
- $x_i \leq u_i, \quad i = 1, \ldots, 4$
  Upper bounds for the releases
- $l_0 \leq x_0 \leq u_0$
  Upper and lower bound for the required capacity of reservoir V

Our objective, as stated above, will be to minimize x_0. All variables must be nonnegative.

The model given above contains one joint chance constraint and eight separate chance constraints. We will now rewrite this model in a way such that only the joint chance constraint still exists explicitly, and the separate chance constraints are already reformulated as linear constraints for the decision variables.

We suppose the reservoir storage to be at its minimum after the vegetation period (period 4), so we have $s_0 = m_4$. We find:

$$s_i = m_4 + \zeta_i - \sum_{j=1}^{i} x_i, \quad i = 1, \ldots, 4$$

We can substitute this into the separate chance constraints to get the following constraints depending on quantiles of the distribution of the random variables ζ_i, denoted by $z_i(\cdot)$:

- $\sum_{j=1}^{i} x_j \leq z_i(1 - \alpha_i) + m_4 - m_i, \quad i = 1, \ldots, 4$
- $\sum_{j=0}^{i} x_j \geq z_i(\gamma_i) + m_4 + v_i, \quad i = 1, \ldots, 4$

The distributions of the random variables are normal, with the coefficients shown in Table 8.5. The values of other variables can be found in Table 8.6.

Table 8.5 Mean and standard deviation of the random variables; correlations of $(\beta_2, \beta_3, \beta_4)$

Period i	$E(\zeta_i)$	$\sigma(\zeta_i)$	$E(\beta_i)$	$\sigma(\beta_i)$		ρ_{ij}	
1	303.47	122.28					
2	375.94	133.43	20.2	8.61	1.0	0.360	0.125
3	432.61	140.27	27.37	10.65	0.360	1.0	0.571
4	486.26	158.64	10.65	6.00	0.125	0.571	1.0

Table 8.6 Values of constants and time-dependent variables

Coefficient	Value
m_i	57 $\forall i$
v_i	70 $\forall i$
α_i	0.95 $\forall i$
γ_i	0.75 $\forall i$

Period	l_i	u_i	d_i
0	100	500.0	–
1	0	102.3	38.1
2	0	252.0	12.5
3	0	252.0	12.5
4	0	252.0	12.5

(a) Solve the chance-constrained model described above ($\alpha = 0.9$). Compute also the reliability of your solution.

(b) Comment on the solution that you found under (a). Hint: which individual chance constraints strongly influence the value of x_0? Analyze the sensitivity of the problem with respect to relevant γ_i and α_i, $i \in \{1, \ldots, 4\}$.

(c) Formulate and solve a recourse model in which all the above mentioned constraints are taken into account, except the joint chance constraint, with recourse cost being made if the total release is not enough to cover the demand for irrigation one of the periods. That is, the expected recourse is

$$c \cdot E\left\{\max_{l=2,3,4} (\beta_i + d_i - x_i)^+\right\}$$

for a suitable penalty parameter c. (In the paper, $c = 100$ is suggested.) Also compute the Expected Value solution and its Expected Result. Give an interpretation of the outcomes. In particular, discuss the usefulness of solving the recourse model.

(d) Formulate a recourse model in which also not meeting the minimum storage and freeboard constraints is penalized, rather than formulated via chance constraints. Choose suitable penalty parameters and solve the recourse model. Compute the Expected Value solution for this model and its Expected Result. Give an interpretation of the outcomes.

(e) Discuss the results of (a), (c) and (d).

8.3 Budgeting Cost of Nursing in a Hospital[3]

Nursing constitutes the largest single cost element in most hospitals today. In this case we will look at a model for preparing the yearly nursing work force budget.

In our model we will make a distinction between three different classes of nurses. The highest in rank is the *registered nurse*, denoted by RN. The lowest in rank is the *nurse aide* (NA). Between these two stands the *licensed vocational nurse* (LVN). We will assume that the work force level decided upon before the budget lead time

[3] See Kao and Queyranne [14].

cannot be changed, although it is possible to hire temporary help or work overtime. The demand for nursing hours is of a stochastic type.

In the following we will look at a recourse model representation of this problem, where our objective will be to minimize total expected cost.

For each nursing skill class, overtime hours are more expensive than regular hours, and hiring from agencies is still more expensive. On the other hand, the regular workforce is fixed during the year whereas overtime hours (up to a maximum percentage of the regular time) and hired hours can be used in a flexible way, depending on the actual needs.

Of course, RN hours are more expensive than LVN hours, and LVN hours are more expensive than NA hours. The reason is, that nurses of a higher skill class are entitled to do more services than nurses of a lower skill class. In the model to be formulated the demand for nursing services is aggregated over the skill classes. Without additional constraints cost minimization would lead to the use of only NA nurses. In order to make the model more realistic, constraints will be used guaranteeing that sufficient higher skill class hours are available.
We define the following parameters and variables:

t	Index of successive months, $t = 1, \ldots, L$
L	Budget lead time. For practical reasons, we take $L = 6$ rather than $L = 12$
i	Index of nursing skill classes; $i = 1, 2, 3$ for RN, LVN, NA respectively
R_i	Number of regular-time nursing hours in skill class i, to be allocated in each month t
$\mathcal{O}_{it}$	Hours of overtime for nursing class i, in month t
A_{it}	Hours of hired nursing, skill class i, in month t
d_t	Random demand in nursing hours, in month t
p_t	Fraction of the regular-time work force that is productive in month t, as a result of holidays and sick leaves. This fraction is assumed constant over all nursing classes
$\tilde{r}_i$	Average cost per hour of regular time in skill class i, in dollars
o_i	Average cost per hour of overtime in skill class i, in dollars
a_i	Average cost per hour of hired time in skill class i, in dollars
r_i	Average cost of regular time in skill class i per hour during L months, in dollars ($r_i = \tilde{r}_i L$)
g	The maximum available overtime in any month and for any skill class is bounded by a factor g times the productive regular-time work force of that skill class in that month
b_i	The number of nurses in skill class i is not allowed to exceed b_{i-1} times the number of nurses in skill class $i - 1$ ($i = 2, 3$)

In our recourse model the first-stage decisions will deal with the number of regular nurse hours in each month in each skill class, while the second-stage decisions will deal with the amount of overtime and the number of nurse hours to be hired, in order to satisfy demand. The second-stage constraints will look like this:

- $\sum_{i=1}^{3}(p_t R_i + \mathcal{O}_{it} + A_{it}) \geq d_t, \quad t = 1, \ldots, L$
  Total work force must be sufficient to satisfy the demand in each month.
- $\mathcal{O}_{it} \leq g \cdot p_t R_i, \quad i = 1, 2, 3, \ t = 1, \ldots, L$
  Upper bounds on the available overtime.
- $p_t R_i + \mathcal{O}_{it} + A_{it} \leq b_{i-1}(p_t R_{i-1} + \mathcal{O}_{i-1,t} + A_{i-1,t}), \quad i = 2, 3; t = 1, \ldots, L$
  In each month, the total number of nursing hours in class $i-1$ must be at least the total number of nursing hours in class i divided by b_{i-1}.

The recourse problem now looks as follows:

$$\min_R \left\{ \sum_{i=1}^{3} r_i R_i + \sum_{t=1}^{L} \mathbb{E}_{d_t}\left[c_t(R, d_t)\right] : R = (R_1, R_2, R_3) \geq 0 \right\}$$

with

$$c_t(R, d_t) = \min_{\mathcal{O},A} \left\{ \sum_{i=1}^{3}(o_i \mathcal{O}_{it} + a_i A_{it}) : \begin{array}{l} \mathcal{O} = (\mathcal{O}_{it}) \geq 0, \ A = (A_{it}) \geq 0, \\ \text{constraints} \end{array} \right\}.$$

The random variables d_t, $t = 1, \ldots, L$, follow a normal distribution with the parameters shown in Table 8.7. The values of the parameters p_t are also shown in this table. The values of the unit costs and the remaining parameters are given in Table 8.8.

In SLP-IOR notation the second-stage optimization problem is

$$\min\{qy : Wy \geq h - Tx, \ y \in \mathbb{R}_+^{36}\}$$

where $q = (q_1, q_2, \ldots, q_6)$,

$$q_t = (o_1, o_2, o_3, a_1, a_2, a_3),$$

Table 8.7 Distribution of demand in hours and productivity fraction, per month

Month t	m_t	σ_t	p_t
1	11975	1637	0.8943
2	11740	1621	0.8917
3	12169	1652	0.8948
4	13132	1714	0.9086
5	13525	1740	0.9032
6	12598	1680	0.8842

Table 8.8 Average cost per hour (in US$)

Nursing class i	$\tilde{r}_i$	o_i	a_i	b_i	g
1	7.03	9.59	11.70	0.6	–
2	4.53	6.18	9.95	2.0	–
3	3.44	4.69	5.78	–	
	–	–	–	–	0.2

$y = (y_1, y_2, \dots, y_6)',$

$$y_t = (\mathcal{O}_{1t}, \mathcal{O}_{2t}, \mathcal{O}_{3t}, A_{1t}, A_{2t}, A_{3t})',$$

$W = \text{diag}(W_1, W_2, \dots, W_6),$

$$W_t = \begin{pmatrix} 1 & 1 & 1 & 1 & 1 & 1 \\ -1 & 0 & 0 & 0 & 0 & 0 \\ 0 & -1 & 0 & 0 & 0 & 0 \\ 0 & 0 & -1 & 0 & 0 & 0 \\ b_1 & -1 & 0 & b_1 & -1 & 0 \\ 0 & b_2 & -1 & 0 & b_2 & -1 \end{pmatrix},$$

$T = (T_1, T_2, \dots, T_6)',$

$$T_t = \begin{pmatrix} p_t & p_t & p_t \\ p_t g & 0 & 0 \\ 0 & p_t g & 0 \\ 0 & 0 & p_t g \\ p_t b_1 & -p_t & 0 \\ 0 & p_t b_2 & -p_t \end{pmatrix},$$

$x = (R_1, R_2, R_3)',$
$h = (h_1, h_2, \dots, h_6)',$

$$h_t = (d_t, 0, 0, 0, 0, 0)'.$$

(a) We will analyze the recourse model described above, with $L = 6$.
Compute (i) the Expected Value Solution, and (ii) the Expected Result (of the expected value solution). Give an interpretation of the outcomes. In particular, discuss the usefulness of solving the recourse model.
Solve the recourse model and compare the outcome with (i) and (ii) above. In particular, discuss the so-called Value of the Stochastic Solution.
Compute the Wait-and-See solution and compare the outcome with previous results. In particular, discuss the so-called Expected Value of Perfect Information.

(b) (Optional) Analyze the sensitivity of the problem with respect to $\tilde{r}_i$, o_i and a_i, $i = 1, 2, 3$. Look at one or more realistic scenarios, for example what happens when the wages go up, so when $\tilde{r}_i$, o_i and a_i rise.

(c) It is clear that the size of the second-stage problem is rather large due to the fact that different months are dealt with separately, and the nursing hours have been disaggregated in skill classes. In this part of the exercise you are asked to formulate and solve a simplified model, and to discuss the results.

Hints

1. Take $L = 1$, and use averaged values for the distribution of the demand $\bar{d}$ and the productivity fraction $\bar{p}$.
2. Assign *fixed* fractions μ_i of the total regular hours (overwork hours, hired hours) to skill class i. That is, with R ($\mathcal{O}$, A respectively) the total number of regular hours in any month (overwork hours, hired hours, respectively), set

$$R_i = \mu_i R,\ \mathcal{O}_i = \mu_i \mathcal{O},\ A_i = \mu_i A$$

Select values for μ_i, $i = 1, 2, 3$, such that the conditions on the mix of classes are satisfied.

Then the recourse model becomes

$$\min_{R \in \mathbb{R}_+} \{\tilde{r} R + E\, c(R, \bar{d})\}$$

with

$$c(R, \bar{d}) = \min_{(\mathcal{O}, A) \in \mathbb{R}^2_+} \left\{ o\mathcal{O} + aA \ : \ \begin{array}{ll} \mathcal{O} + A & \geq \bar{d} - \bar{p}R \\ \mathcal{O} & \leq g\bar{p}R \end{array} \right\}$$

where $\tilde{r} = \sum_{i=1}^{3} \mu_i \tilde{r}_i$, $o = \sum_{i=1}^{3} \mu_i o_i$, $a = \sum_{i=1}^{3} \mu_i a_i$.

(d) Formulate and analyze a model which reflects the possibility to put patients on a waiting-list. Hint: This can be modelled by a recourse structure which has recourse matrix $(W\ \bar{W})$, where W is the original recourse matrix and the 36×6 matrix $\bar{W}$ has non-zeros in row $6k + 1$, $k = 0, 1, \ldots, 5$, corresponding to, respectively, row $k + 1$ of the following matrix:

$$\begin{pmatrix} 1 & 0 & 0 & 0 & 0 & -1 \\ -1 & 1 & 0 & 0 & 0 & 0 \\ 0 & -1 & 1 & 0 & 0 & 0 \\ 0 & 0 & -1 & 1 & 0 & 0 \\ 0 & 0 & 0 & -1 & 1 & 0 \\ 0 & 0 & 0 & 0 & -1 & 1 \end{pmatrix}$$

Give an interpretation of this recourse structure. In particular, discuss the handling of the planning horizon, and show (using the solution) that treatment is never delayed more than 1 month. Compare the outcome with the results found under (a).

8.4 Growing Maize and Sorghum in Kilosa District[4]

Consider a family in the Kilosa District, Tanzania. This family has some land on which it can grow crops to satisfy its own needs. In this case study we distinguish only two crops, maize and sorghum. The yields of the crops are uncertain, partly because of uncertain rainfall conditions. Insufficient food production may lead to malnutrition, i.e. shortage of nutrients. We will consider only two nutrients, calories and protein. We have the following data:

- the family needs $44 \cdot 10^5$ Kcal and 89 kg of proteins per year
- 100 g of maize meal contain 354 Kcal and 8 g of proteins
- 100 g of sorghum meal contain 355 Kcal and 10 g of proteins
- circa 20% of the yields of both maize and sorghum are lost when they are ground to flour

From this we see that:

- 100 kg of maize contain $2.8 \cdot 10^5$ Kcal and 6.4 kg of proteins for consumption
- 100 kg of sorghum contain $2.8 \cdot 10^5$ Kcal and 8.0 kg of proteins for consumption

If we look at these numbers, we could conclude that growing sorghum would be more rewarding, in terms of nutritional value, than growing maize, since the amount of calories obtained is equal for both crops, whereas the amount of proteins is larger for sorghum. However, nutritional value is not the only aspect: taste is important too. Maize is much preferred over sorghum, so our family would rather have maize than sorghum, if possible. We will make this preference explicit by assuming that the family wants to eat, say, at least three times as much maize as sorghum. In the following we will look at a (complete) recourse model describing this situation. The first-stage decisions deal with the number of ha assigned to maize and sorghum, respectively. The second-stage deals with buying of maize and/or sorghum, in case the yields are insufficient. Since labor is relatively inexpensive, and money is scarce, the objective is to minimize the expected cost of buying maize and sorghum afterwards. If the yields happen to be very good, it is assumed that surpluses are sold. However, the selling prices are (much) lower that the buying prices in the model, since it is expected that in a good year prices are low, and in a bad year they will be high.

---

[4]See Schweigman [43] pp. 118–123, 138–141.

We define the following variables:

x_1	Acreage of maize, in ha
x_2	Acreage of sorghum, in ha
m^+, m^-	Shortage/surplus of maize, to be bought/sold, in units of 100 kg
s^+, s^-	Shortage/surplus of sorghum, to be bought/sold, in units of 100 kg
Y_1	Random yield per ha of maize, in units of 100 kg
Y_2	Random yield per ha of sorghum, in units of 100 kg
R	Random total rainfall during the growing season, in mm
ε_i	Random disturbances, $i = 1, 2$
f	Preference factor of maize with respect to sorghum
a	Available acreage, in ha
q_m^+, q_m^-	Buying/selling price of maize, in euros per 100 kg
q_s^+, q_s^-	Buying/selling price of sorghum, in euros per 100 kg

Regression analysis applied to empirical data leads to the following model for the random yields.

The yields Y_i, $i = 1, 2$, (in units of 100 kg/ha) depend on total rainfall (R) and disturbances ε_i:

$$Y_1 = 0.020\, R - 1.65 + \varepsilon_1$$

$$Y_2 = 0.008\, R + 5.92 + \varepsilon_2$$

where R and ε_1, ε_2 have the following distributions

$$R \sim \mathcal{N}(\mu = 515.5,\ \sigma^2 = 18,769)$$

$$\varepsilon_1 \sim \mathcal{N}(\mu = 0,\ \sigma^2 = 100)$$

$$\varepsilon_2 \sim \mathcal{N}(\mu = 0,\ \sigma^2 = 100)$$

The second-stage constraints will be:

- $2.8(Y_1 x_1 + m^+ - m^-) + 2.8(Y_2 x_2 + s^+ - s^-) \geq 44$,
  The amount of calories has to be sufficient (expressed in 10^5 KCal).
- $6.4(Y_1 x_1 + m^+ - m^-) + 8.0(Y_2 x_2 + s^+ - s^-) \geq 89$,
  The amount of proteins has to be sufficient (expressed in kg protein).
- $(Y_1 x_1 + m^+ - m^-) - f(Y_2 x_2 + s^+ - s^-) \geq 0$,
  The family wants to eat at least f times as much maize as sorghum (expressed in units of 100 kg).

We have only one first-stage constraint:

- $x_1 + x_2 \leq a,$
  The acreage available is bounded.

The recourse problem now looks as follows:

$$\min \{Q(x_1, x_2) : x_1 + x_2 \leq a,\ x_1, x_2 \geq 0\}$$

where

$$Q(x_1, x_2) = \mathbb{E}_{Y_1, Y_2}\left[v(x_1, x_2, Y_1, Y_2)\right]$$

and

$$v(x_1, x_2, Y_1, Y_2) =$$
$$\min \left\{ q_m^+ m^+ - q_m^- m^- + q_s^+ s^+ - q_s^- s^- : m^+, m^-, s^+, s^- \geq 0 \right\}$$

and subject to the restrictions described above.

(a) Prove that this is a complete recourse model. (And yes, it *is* complete.)

(b) Assume that the buying price of maize is 50 euros per 100 kg; the buying price of sorghum is 40 euros per 100 kg; the selling price of maize is 35 euros per 100 kg; the selling price of sorghum is 30 euros per 100 kg; total acreage $a =$ 1.5 ha; the preference factor $f = 3$. Assume that the disturbances ε_1 and ε_2 are independent random variables, and that they are independent of the total rainfall R.
Compute (i) the Expected Value Solution, and (ii) the Expected Result (of the expected value solution). Give an interpretation of the outcomes. In particular, discuss the usefulness of solving the recourse model.
Solve the recourse model and compare the outcome with (i) and (ii) above. In particular, discuss the so-called Value of the Stochastic Solution.
Compute the Wait-and-See solution and compare the outcome with previous results. In particular, discuss the so-called Expected Value of Perfect Information.

(c) Analyze the sensitivity of the problem with respect to the parameters a and f and the assumption of independence of ε_1 and ε_2. How does the solution change if the buying prices are doubled, and the selling prices become zero?

(d) Assume that the prices $q_m^+, q_m^-, q_s^+, q_s^-$ are (normally distributed) random variables. Choose reasonable correlations with the other random variables and analyze the model. Compare with (b).
Hints: (i) To avoid that e.g. $q_m^+ < -q_m^-$ for some realizations (so that the problem is unbounded), use a random variable q_m and define e.g. $q_m^+ = q_m$ and $q_m^- = -0.7q_m$.

(ii) For the Kilosa District, a reasonable correlation between total rainfall and prices is −0.7.

(e) (Optional) Suppose that there is no possibility of trading maize and sorghum. On the other hand, there is a possibility to use more land. Suppose also that a ha sorghum asks for similar workload as a ha maize. Then it is important to calculate the minimum acreage such that the family is selfsufficient with a reasonable reliability. This leads to the following model:

$$\begin{aligned} \min\ & x_1 + x_2 \\ \text{s.t. } & \Pr\{2.8Y_1x_1 + 2.8Y_2x_2 \geq 44\} \geq \alpha_1 \\ & \Pr\{6.4Y_1x_1 + 8.0Y_2x_2 \geq 89\} \geq \alpha_2 \\ & x_1 \geq 0,\ x_2 \geq 0 \end{aligned}$$

It is possible to reduce the constraints to linear and quadratic constraints. Do this and solve the problem. Compare the results with (b).

8.5 Product Mix Problem[5]

A furniture shop has 6000 man-hours available in the carpentry shop and 4000 man-hours in the finishing shop per year. These are the hours that the permanent staff is supposed to work in these shops. The actual man-hours available, however, are assumed to be normally distributed random variables with deficits resulting from employee absences and surpluses due to reduction of workload elsewhere. There are four classes of products each consuming a certain number of man-hours in carpentry and finishing; the actual time consumed is assumed to be a uniformly distributed random variable. Each product earns a certain profit per item, and the shop has the option to purchase casual labor from outside. The problem is to find a product mix, such that the permanent employees are used as profitable as possible, taking into account the uncertainties specified above. Note that the cost of the salaried labor is fixed, and thus does not enter the problem.

In the following we will look at a simple recourse representation of this problem. Its indices, parameters, and variables are described in the list on the next page.

As objective we choose to maximize expected profit. Since most solvers in SLP-IOR do not accept maximization problems, we formulate the model as a minimization problem:

$$\min_{x \geq 0} -\sum_{j=1}^{4} c_j x_j + Q(x)$$

[5]See Ermoliev and Wets [6] p. 554–555.

j	Product class index, $j = 1, 2, 3, 4$
k	Shop index: $k = 1$ carpentry shop, $k = 2$ finishing shop
x_j	Number of units of product class j to be produced per year
v_k^+	Actual number of hours of casual labor required in shop k per year
v_k^-	Actual number of surplus man-hours in shop k per year
t_{kj}	Random number of hours required in shop k to produce one unit of product class j
h_k	Random number of hours of salaried labor available per year in shop k
c_j	Profit per unit of product class j
q_k	Cost per hour of casual labor in shop k

with

$$Q(x) = \mathbb{E}_{h,t}\left[\min_{v^+,v^-\geq 0}\left\{\sum_{k=1}^{2} q_k v_k^+ \;:\; v_k^+ - v_k^- = \sum_{j=1}^{4} t_{kj}x_j - h_k,\ k = 1, 2\right\}\right].$$

The distributions of the random variables are given in Table 8.9. They are assumed to be independent. (Is this relevant?) The profits c_j per unit are 12, 20, 18 and 40 respectively, and the cost q_k per hour for casual labor are 5 and 10 respectively, all in dollars.

(a) Replace all distributions by 'lower bounding' discrete distributions on two points.
Compute (i) the Expected Value Solution, and (ii) the Expected Result (of the expected value solution). Give an interpretation of the outcomes. In particular, discuss the usefulness of solving the recourse model.

Table 8.9 Distributions in the product mix problem

Variable	Distribution
h_1	$\mathcal{N}\ (6000, \sigma^2 = 10,000)$
h_2	$\mathcal{N}\ (4000, \sigma^2 = 2500)$
t_{11}	$\mathcal{U}\ (3.5, 4.5)$
t_{12}	$\mathcal{U}\ (8, 10)$
t_{13}	$\mathcal{U}\ (6, 8)$
t_{14}	$\mathcal{U}\ (9, 11)$
t_{21}	$\mathcal{U}\ (0.8, 1.2)$
t_{22}	$\mathcal{U}\ (0.8, 1.2)$
t_{23}	$\mathcal{U}\ (2.5, 3.5)$
t_{24}	$\mathcal{U}\ (36, 44)$

Solve the recourse model and compare the outcome with (i) and (ii) above. In particular, discuss the so-called Value of the Stochastic Solution.
Compute the Wait-and-See solution and compare the outcome with previous results. In particular, discuss the so-called Expected Value of Perfect Information.

(b) Replace all distributions by 'upper bounding' discrete distributions on two points. Analyze the problem.
(c) Analyze the recourse model with the original continuous distributions.
(d) Compare the results of (a), (b) and (c).
(e) Analyze the sensitivity of the problem under (a) with respect to the discretization. Do this for example by varying the choice of the underlying intervals.

8.6 Investment Planning for Electricity Generation[6]

The demand for electricity fluctuates heavily with time. During peak-hours much more is needed than in quiet times, and in winter the demand is higher than in summer. Electricity producers usually represent the demand over a prespecified period of time (a year, say) in terms of the so-called *load-curve*, see Fig. 8.1.

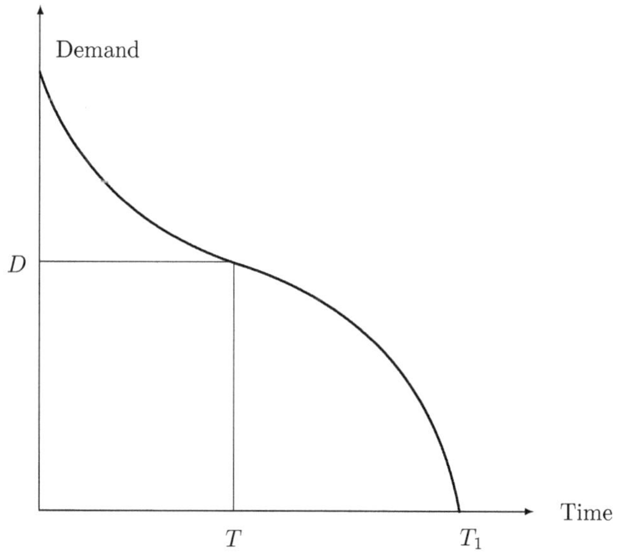

Fig. 8.1 Load-curve, $D = L(T), 0 \leq T \leq T_1$

[6]See Louveaux and Smeers [24].

Here T_1 indicates the number of time units (a day, say) in the considered period. The load-curve is constructed by representing the days in the order of decreasing demands. As a result, the load-curve is a nonincreasing function of T on $[0, T_1]$. Note that the load-curve does not reveal on which days the corresponding demand occurs, only the frequency is depicted. That is, $D = L(T)$ means, that in the period of T_1 time units the demand is at least D for in total T time units.

The load-curve of a year is a practical means of representing the variability of the demand over time, in particular for analyzing the technology to be used to generate electricity, as is the subject of this case. On an aggregated level, one may think of production technologies based on coal, natural gas, sun, wind, nuclear power and so on. Each technology has its own advantages and disadvantages. From the point of view of costs, equipment with relatively low operating cost will be suitable, even if the investment costs are large, if it is exploited during a large part of the year, where it may be the other way around for peak-load demand.

In the article of Louveaux and Smeers (on which this case is based) a multistage stochastic programming model is formulated for the investment planning for electricity generation. In this case we restrict ourselves to a two-stage model with recourse. The first stage will deal with decisions on investments in the technologies, whereas the second stage deals with the distribution of the technologies over the various parts of the load-curve. Since there is a long lead-time for the investments, the load-curve is not known exactly in advance, and will be modeled in terms of random variables, see Fig. 8.2.

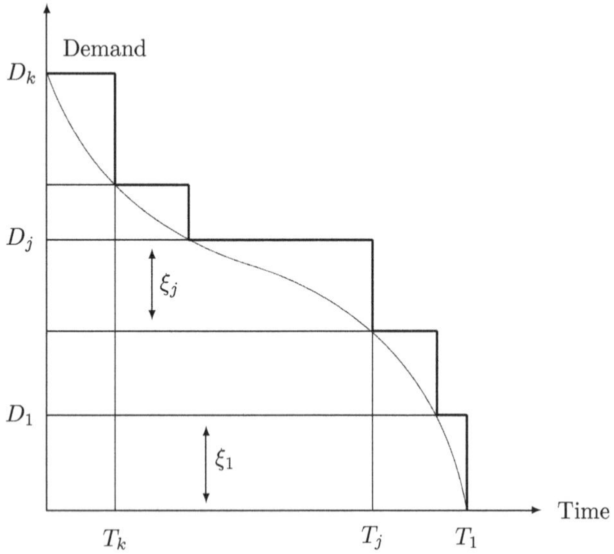

Fig. 8.2 Approximate load curve, with k modes. The random demand in mode j is $\xi_j := D_j - D_{j-1}$, $j = 1, \ldots, k$ with $D_0 := 0$

The approximate load-curve is a piecewise constant curve, indicating that during T_j time units the demand is equal to D_j, $j = 1, \ldots, k$. In our model we will assume that the values of $T_1, \ldots, T_k$ are known whereas $D_1, \ldots, D_k$ are random variables, specified as $D_h = \sum_{j=1}^{h} \xi_j$, $h = 1, \ldots, k$. That is, ξ_j is the demand related with the j-th block in the approximate load-curve, to be called the j-th "mode". For simplicity, we will assume that there does not exist any capacity in any technology in the beginning of the investment period.

Define the following variables.

n	Number of technologies
k	Number of modes
x_i	Capacity for technology i, $i = 1, \ldots, n$
y_{ij}	Capacity of technology i effectively used in mode j, $i = 1, \ldots, n$, $j = 1, \ldots, k$
ξ_j	(Stochastic) demand for mode j, $j = 1, \ldots, k$
$\xi_j^{\max}$	(Deterministic) maximum possible value for ξ_j, $j = 1, \ldots, k$
T_j	Duration of mode j, $j = 1, \ldots, k$
c_i	Investment and maintenance cost for technology i per unit capacity, $i = 1, \ldots, n$, (on a yearly equivalent basis)
q_i	Production cost for technology i per unit capacity per unit of time, $i = 1, \ldots, n$
$c_{\max}$	Maximum allowed investment cost

The first-stage constraint will look like this:

- $\sum_{i=1}^{n} x_i \geq \sum_{j=1}^{k} \xi_j^{\max}$
  The total available capacity has to be enough to cover maximum demand
- $\sum_{i=1}^{n} c_i x_i \leq c_{\max}$
  Budget constraint

The second-stage constraints look like this:

- $\sum_{j=1}^{k} y_{ij} \leq x_i, \quad i = 1, \ldots, n$
  The modes cannot use more capacity than is available for any technology
- $\sum_{i=1}^{n} y_{ij} \geq \xi_j, \quad j = 1, \ldots, k$
  Demand has to be met in each mode

Our objective is to minimize expected total cost, which is represented by the following formula:

$$\min_{x \geq 0} \left\{ \sum_{i=1}^{n} c_i x_i + \mathbb{E}_\xi \left[\min_{y \geq 0} \left\{ \sum_{i=1}^{n} \sum_{j=1}^{k} q_i T_j y_{ij} : \begin{array}{c} \text{sec.-stage} \\ \text{constr.} \end{array} \right\} \right] : \begin{array}{c} \text{first-stage} \\ \text{constr.} \end{array} \right\}$$

We will assume that four technologies are available, which can all be operated in three different modes. We assume that $T_1 = 10$, $T_2 = 0.6 \cdot T_1$ and $T_3 = 0.1 \cdot T_1$. The upper bound on the investment costs is equal to 120, while the investment costs per unit for the separate technologies are 10, 7, 16 and 6, respectively. Production costs are 4.0, 4.5, 3.2 and 5.5, respectively, per unit.

(a) We will analyze the recourse model described above.
Assume that the distributions of the *independent* random variables ξ_i are:

Prob	ξ_1	ξ_2	ξ_3	
0.3	3	2	1	
0.4	5	3	2	(altogether 27 realizations)
0.3	7	4	3	

Compute (i) the Expected Value Solution, and (ii) the Expected Result (of the expected value solution). Based on the results, discuss the usefulness of solving the recourse model. Remark: To compute (i), the maximum demand in the first-stage constraint can be adjusted.
Solve the recourse model and compare the outcome with (i) and (ii) above. In particular, discuss the so-called Value of the Stochastic Solution.
Compute the Wait-and-See solution and compare the outcome with previous results. In particular, discuss the so-called Expected Value of Perfect Information.

(b) Assume now, in addition, that the operational availability of each of the technologies is random: if a capacity of x_i is installed, the actual amount available is $\alpha_i x_i$, with α_i a random parameter, $i = 1, .., n$.

$$\begin{aligned} \alpha_1 &\sim \mathcal{U}(0.6, 0.9) \\ \alpha_2 &\sim \mathcal{U}(0.7, 0.8) \\ \alpha_3 &\sim \mathcal{U}(0.5, 0.8) \\ \alpha_4 &\sim \mathcal{U}(0.9, 1) \end{aligned}$$

Moreover, assume that it is possible to 'import' electricity to cover any observed power shortage. This virtual fifth technology has zero investment cost, but high production cost 10.
Formulate the corresponding recourse model. (Hint: the possibility to import electricity has implications for both stages of the model.)

Analyze this model, and compare the results with (a). In particular, discuss the expected result (in model (b)) of the solution obtained under (a).

(c) Assume, in addition, that $T_i = \tau_i, \ i = 2, 3$, where the distribution of the independent random variables τ_i is

Prob	τ_2	τ_3
0.6	5	0.5
0.4	7.5	1.75

(altogether 4 realizations)

Formulate this model and analyze it. Compare the results with the other models.

8.7 Optimizing the Timetable of the Haarlem–Maastricht Railway Connection[7]

Consider a railway line with a number of stations. To avoid delays, the travelling time of a train between two stations is generally planned to be longer than what is expected in an ideal situation. The differences between the planned and ideal (undisturbed) travelling times between the stations are called *running time supplements*. These are meant to compensate for uncertain disturbances that may result in excess travelling times. (The components of the total travelling time between two stations are graphically represented in Fig. 8.3.)

In fact, almost all delays could be eliminated by simply including a sufficient amount of running time supplements in the timetable. However, the passengers would not favour the resulting excessively long travelling times. Thus, in practice,

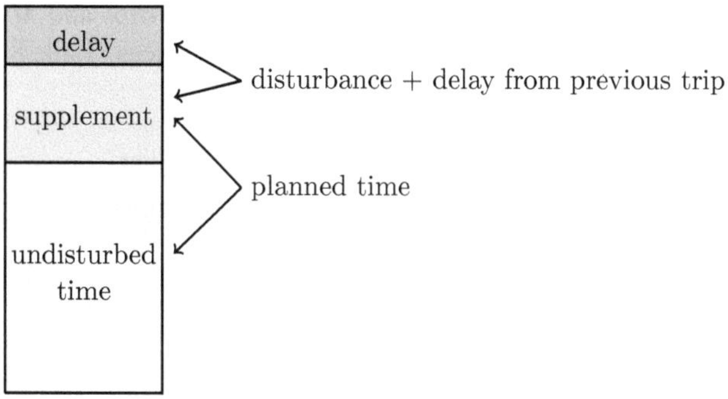

Fig. 8.3 Decomposition of the actual running time on a trip in case of a delay

[7]For further information about this case, see P. Vekas, *Optimizing existing railway timetables and insuring the risk of ticket refund guarantees*, MSc Thesis, Corvinus University, Budapest, 2007.

n	Number of trips on the railway line (i.e., there are $n+1$ stations, see Fig. 8.4)
ω_i $(i = 1, 2, \ldots, n)$	Random disturbance on trip i
x_i $(i = 1, 2, \ldots, n)$	Running time supplement on trip i
d_i	Delay of the train at the end of trip i
M	Total amount of supplements

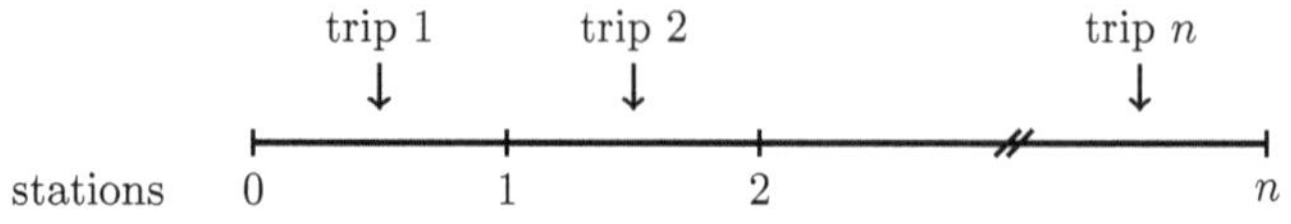

Fig. 8.4 Stations and trips in the model

railway companies restrict the total amount of supplements to some reasonable extent and try to distribute it in the timetable in an efficient way. It is important to notice that this problem is not about inventing a completely new timetable, but rather about finely adjusting an already existing and well-functioning one by minor reallocations.

In the following we will look at a recourse model representing this problem. We define the following variables (time quantities are expressed in minutes):

In the model, we will assume that the train departs from station 0 with no delay and it does not spend any time waiting at the stations (this assumption could be relaxed easily by introducing the periods spent at the stations as additional trips, however, we will not do so for the sake of simplicity). The disturbances ω_i $(i = 1, 2, \ldots, n)$ are assumed to be independent and exponentially distributed with different means. We will assume that there are no negative delays, which is equivalent to saying that the train stays waiting at a station if it accidentally arrives there too early.

The Haarlem–Maastricht railway line contains $n = 8$ trips and $M = 10.93$ on this line. A study has estimated the means of the disturbances (Table 8.10):

Table 8.10 Mean disturbances on the Haarlem–Maastricht railway line

Trip	Mean disturbance (minutes)
Haarlem–Amsterdam Centraal	1.03
Amsterdam Centraal–Duivendrecht	0.84
Duivendrecht–Utrecht Centraal	1.15
Utrecht Centraal–'s Hertogenbosch	2.01
's Hertogenbosch–Eindhoven	1.28
Eindhoven–Roermond	2.40
Roermond–Sittard	1.22
Sittard–Maastricht	0.87

As the goal of the railway company is to minimize the sum of expected delays, the problem may be formulated as follows:

$$\min_{x\geq 0} \{E_\omega[v(x,\omega)] : x_1 + x_2 + \cdots + x_n \leq M\}, \tag{8.1}$$

$$\text{where} \quad v(x,\omega) = d_1(x,\omega) + d_2(x,\omega) + \ldots + d_n(x,\omega).$$

This is not a two-stage recourse model, however, it may be reformulated that way. To accomplish this, first note that x_1 is meant to compensate for ω_1, so $d_1 = (\omega_1 - x_1)^+$, and in general, x_i is meant to compensate for both d_{i-1} and ω_i, so $d_i = (d_{i-1} + \omega_i - x_i)^+$ for all $i = 1, 2, \ldots, n$ with $d_0 \equiv 0$. The second-stage problem that turns (8.1) into a two-stage recourse model is

$$\begin{aligned} v(x,\omega) = \min_{y\geq 0}\{y_1 + y_2 + \ldots + y_n :& \\ y_1 &\geq \omega_1 - x_1, \\ y_2 &\geq y_1 + \omega_2 - x_2, \\ &\vdots \\ y_n &\geq y_{n-1} + \omega_n - x_n, \\ y_1, y_2, \ldots, y_n &\geq 0\}.. \end{aligned} \tag{8.2}$$

This may be verified in the following way: first it may be proved by induction that for every feasible solution of (8.2), $y_i \geq d_i$ $(i = 1, 2, \ldots, n)$, and additionally, it is obvious that $y_i = d_i$ $(i = 1, 2, \ldots, n)$ is a feasible solution of (8.2), and finally, it is not difficult to see that the objective function of (8.2) is monotone increasing in y_i. These three results together imply that the variables y_i will attain the values of d_i in the optimum for every $i = 1, 2, \ldots, n$.

The structure of (8.2) may be captured by the following matrix:

$$W = \begin{pmatrix} 1 & 0 & \cdots & 0 & 0 \\ -1 & 1 & \cdots & 0 & 0 \\ \vdots & \vdots & \ddots & \vdots & \vdots \\ 0 & 0 & \cdots & 1 & 0 \\ 0 & 0 & \cdots & -1 & 1 \end{pmatrix}$$

With this notation, a brief two-stage recourse formulation of our model reads as follows:

$$\min_{x\geq 0} \{E_\omega[v(x,\omega)] : \mathbf{1}x \leq M\}, \tag{8.3}$$

$$\text{where} \quad v(x,\omega) = \min_{y\geq 0}\{\mathbf{1}y : x + Wy \geq \omega\}.$$

(a) We will analyze the recourse model described above.
Compute (i) the Expected Value Solution, and (ii) the Expected Result (of the expected value solution). Based on the results, discuss the usefulness of solving the recourse model.
Solve the recourse model and compare the outcome with (i) and (ii) above. In particular, discuss the so-called Value of the Stochastic Solution.
Compute the Wait-and-See solution and compare the outcome with the previous results. In particular, discuss the so-called Expected Value of Perfect Information.
(b) Analyze the sensitivity of the model with respect to the parameter M.
(c) Using the recourse solution, estimate the probability that the delay of the train in Maastricht will be greater than 3 min.
(d) Solve the model using a solution x with all the $x_i, i = 1, 2, \ldots, n$, being proportional to the expected disturbances on the trips. (This is how the values of the supplements are mostly determined in practice.) How does this solution compare to the optimal one?

8.8 Electricity Distribution[8]

A distributor of electricity can be seen as a middleman between suppliers and consumers of electrical power, whose aim is to satisfy demand at minimal costs. We assume that demand for electricity is aggregated, whereas a crucial distinction is made between two categories of suppliers: on the one hand there are power plants (aggregated) and in addition we consider so-called *small generators*. Typical examples of small generators are:

- Hospitals or other institutions with emergency generators, which can supply to the net in normal circumstances;
- Greenhouses, which produce electrical power as a by-product of generating heat;
- Industrial consumers which may be switched off for short periods of time. The net effect is the same as generating an amount of electrical power equal to their demand during such a period.

We consider two planning problems faced by a typical (Dutch) distributor of electricity (in the sequel simply called 'the distributor').

- First, every year a contract with the power plants has to be negotiated. The main issue in this contract is to determine *quota* for the capacity of the power supply. The quota define capacity ranges for supply at any moment during the contract year, and to each range corresponds a certain price per kWh (kilo Watt hour) of supply. For a number of ranges (low, medium, and high volume of supply)

[8] See Klein Haneveld and Van der Vlerk [21].

such quota have to be specified at given costs per kW (kilo Watt) of reserved capacity. These quota are then fixed for the entire year covered by the contract, thus determining the costs of supply by the power plants at every moment in that year. The decision on quota is of course complicated by the uncertainty of future demand, but also by the possibility to use supply by small generators (subject to given contracts).

- Once the quota are fixed, the second problem is to determine a supply schedule (that is, a schedule of supplies obtained from both power plants and small generators for every hour or even every 5 min) for each single day. Such a schedule has to be specified one day in advance. At this time the demand pattern for the next day is known, but only probabilistic information is available on demand in the rest of the contract year. Moreover, such a schedule has to take into account bounds on the total number of hours that a small generator can be used during the contract year.

Small generators can either supply at full capacity or not at all. Thus, the contracts with small generators specify a fixed price for a fixed capacity per time unit and usually also costs for switching them on or off. Moreover, there may be bounds on the number of time periods a small generator can be used, the number of times it can be switched on, up and down times, etc. A natural way to model these contracts is to use *binary* variables.

The uncertainty about future demand is modeled by random variables with known distributions. Because discrete variables are needed to model the contracts with small generators, the models we consider are stochastic programs with mixed-integer variables. In fact, as will be explained below, it is very natural to model both problems as recourse models. We will argue that it is also natural to aggregate the future in each model (consisting of an entire year or the remainder of the year, respectively) and consider it as one time stage, so that both models have a two-stage integer recourse structure.

8.8.1 Contract with Power Plants

For reasons that are beyond the scope of this example, the unit price paid by the distributor for electrical power supplied by the Dutch power plants depends on *quota*. For a given hour t, let this amount of supply be s_t. Then

$$s_t = x_t^1 + x_t^2 + x_t^3 + x_t^4,$$

with

$$\begin{aligned} 0 &\le x_t^1 \le L, \\ 0 &\le x_t^2 \le M, \\ 0 &\le x_t^3 \le H, \\ 0 &\le x_t^4. \end{aligned}$$

That is, the total supply consists of four components, the first three of which are bounded from above by their respective quota L (for Low), M (Middle), and H (High). The corresponding unit prices c^i are such that $c^1 < c^2 < c^3 \ll c^4$, for example $c^1 = 0.04$, $c^2 = 0.05$, $c^3 = 0.09$, and $c^4 = 1.25$.[9] Because of this relation between the prices it is obvious that, for $i = 2, 3, 4$, $x_t^i > 0$ only if x_t^{i-1} equals its upper bound. The interpretation is that a marginal unit of supply s_t becomes more expensive, depending on the classification of the total supply as 'low', 'middle', 'high', or even 'very high'. Supply is considered to be very high if $s_t > L+M+H$; the corresponding marginal unit price c^4 reflects its function as a penalty price.

Clearly, the values of the quota L, M, and H are very important to the distributor, since they determine the costs of the supply by the power plants for every time period. The values of the quota are set in a yearly contract between the distributor and the (collective) power plants, and remain fixed for the entire year covered by this contract. The unit costs (per kW) for reserving capacity are decreasing from L to H, for example 260 for L, 205 for M, and 125 for H. Thus, when deciding on the contract with the power plants, the distributor would like to determine quota that result in the lowest *total expected costs* possible. These costs consist of immediate costs for reserving capacity, and expected future costs for satisfying uncertain future demand during the entire year covered by the contract. This decision under uncertainty is complicated further by the possibility to satisfy a part of the future demand by means of supply from small generators.

8.8.2 Contract with Small Generators

Typically, there are about 30–40 small generators $\{SG^j, j \in J\}$ that can supply to the distributor. In our models the contracts with the small generators are given. In this section we discuss the nature of these contracts, that is, the price structure of supply and the various constraints that may be specified.

As stated in the introduction, a small generator SG^j can either supply at a fixed capacity or be switched off. Hence, the contract contains a fixed price q^j for a fixed amount of supply during one time period (say one hour) and also costs r^j for switching the small generator on or off. Actually, we only model costs for switching on SG^j: observing that SG^j has to be switched off once it has been switched on, we take these costs equal to the sum of both costs.

The decision to use the supply of SG^j during the time period $[t, t+1)$, $t \in \{0, 1, \ldots, 23\} =: T$, is modeled by a binary variable y_t^j, which has value one if the supply is used. If SG^j was off in the previous period $[t-1, t)$ it has to switched on, which is modeled by the binary variable z_t^j which is then set to one: $z_t^j = \max\{y_t^j - y_{t-1}^j, 0\}$.

[9]These numbers are fictitious but the proportions are not unrealistic. The same is true for the values of other cost parameters in this example.

The contract may contain various constraints which restrict the use of SG^j per day. Below we give an overview, together with their possible mathematical formulation.

(i) Supply is not possible during certain hours of the day:

$$\sum_{t \in T_0} y_t^j = 0,$$

where T_0 is the corresponding set of hours.

(ii) SG^j has to be on at least m_1 and at most M_1 hours:

$$m_1 \leq \sum_{t \in T} y_t^j \leq M_1.$$

(iii) SG^j has to be switched on at least m_2 and at most M_2 times:

$$m_2 \leq \sum_{t \in T} z_t^j \leq M_2.$$

(iv) If SG^j is switched on, it has to stay on for at least m_3 and at most M_3 periods: for all $t \in T$ (with appropriate modifications for large t),

$$\sum_{s=t}^{t+m_3-1} y_s^j - m_3 z_t^j \geq 0$$

$$\sum_{s=t}^{t+M_3} y_s^j \quad M_3 \leq 0$$

Below we use the compact notation $(y^j, z^j) \in C^j$ to denote these daily constraints in the contract with SG^j; here y^j is the vector $(y_0^j, \ldots, y_{23}^j)$ and z^j is defined analogously. Of course, a specific contract need not contain all of the constraints mentioned above.

In addition to the daily constraints as described above, there are usually upper and lower bounds on the use of SG^j during the entire contract year. In the model for the contract with the power plants they are ignored or approximated by constraints of type (ii). They will be modeled explicitly in our second model for optimizing the daily supply schedules.

8.8.3 *Modeling Uncertain Demand for Electricity*

The decision on the quota (L, M, H) for next year has to be taken in advance, that is, under uncertainty about the demand for electricity. Not only the daily volume of demand is relevant, but also its dispersion over that day. So the demand for electricity on one day is a vector $\omega = (\omega_0, \omega_1, \ldots, \omega_{23})$, where each component describes the demand in each hour. We assume that a probability distribution for such demand data is available, for example based on historical data.

In order to arrive at a manageable model size, we do not consider random demand for each future day separately, but instead consider so-called *representative days*. An example of a representative day is a working day in the spring season, which represents all working days in that season. This aggregation is motivated by the assumption that all working days in the spring have a similar demand pattern and volume. On the other hand, the demand is different from that on e.g. a holiday or a Saturday in the same season, and also different from the pattern and/or volume observed on working days during other seasons.

Let $R^1, \ldots, R^N$ denote the collection of representative days, for example consisting of working days, Saturdays, Sundays, and holidays for each season, giving $N = 16$. From the available data now a discrete probability distribution can be estimated for the representative days. To simplify the notation in the model to be described, we will assume that all 365 days of the year have the same distribution of the demand pattern ω.

8.8.4 Model 1: Optimizing the Contract with the Power Plants

We now will formulate a model for the first problem of the distributor, as indicated before. Here the quota (L, M, H) are the first-stage decisions, since they are taken before the uncertain demands are realized. The entire year covered by the contract is aggregated to one single stage. So we get a two-stage stochastic model, where in the second stage all 365 days are dealt with separately. For given quota (L, M, H) and a realization of a one-day demand pattern ω, the minimal costs for satisfying this demand are

$$
\begin{aligned}
v(L, M, H, \omega) = \min_{x,y,z} & \sum_{t \in T} \left(\sum_{i=1}^{4} c^i x_t^i + \sum_{j \in J} \left(q^j y_t^j + r^j z_t^j \right) \right) \\
\text{s.t.} \quad & \sum_{i=1}^{4} x_t^i + \sum_{j \in J} b^j y_t^j \geq \omega_t, && t \in T \\
& x_t^1 \leq L, \; x_t^2 \leq M, \; x_t^3 \leq H, && t \in T \\
& (y^j, z^j) \in C^j, && j \in J \\
& x_t^i \geq 0, && i = 1, \ldots, 4, \; t \in T \\
& y_t^j, z_t^j \in \{0, 1\}, && j \in J, \; t \in T
\end{aligned}
$$

where J is the set of small generators. The first constraint reflects that demand has to be satisfied at all times, either by supply x_t^i from the power plants or by supply from small generators, which have fixed capacity b^j, $j \in J$. This is a mixed-integer (binary) optimization problem, with $4 \cdot |T|$ continuous variables and $2 \cdot |J| \cdot |T|$ binary variables. (Notation: $|S|$ denotes the cardinality of a set S.)

Thus, for fixed quota (L, M, H), the expected costs for satisfying the demand of one day are $\mathbb{E}_\omega [v(L, M, H, \omega)]$, giving yearly expected costs $365 \cdot \mathbb{E}_\omega [v(L, M, H, \omega)]$, say. (Of course, if the differences between the demand pattern distributions are taken seriously, here 365 (or less, using representative days) different expected daily costs should be calculated.) In the contract with the power plants, the goal of the distributor is to determine quota that minimize the *total expected costs*, which is the sum of the direct cost for fixing the quota and the expected costs of satisfying demand during the contract year. Thus, the goal is to find an optimal solution of the following *two-stage integer recourse* problem:

$$\begin{aligned} \min_{L,M,H} \quad & C^1 L + C^2 M + C^3 H + 365 \cdot \mathbb{E}_\omega [v(L, M, H, \omega)] \\ \text{s.t.} \quad & L, M, H \geq 0, \end{aligned}$$

where C^i, $i = 1, 2, 3$, are the unit costs for reserving capacity by means of the quota, and v is the second-stage value function defined above. The first-stage decisions are continuous, whereas the second stage contains many *binary decisions*.

8.8.5 Model 2: Supply Schedule for the Next Day

The second problem faced by the distributor is to determine a schedule for the next day, specifying the amount of supply from the power plants and the small generators for each period during that day. This schedule has to be conveyed to the suppliers one day ahead (at noon).

The goal of the distributor is to find a schedule that minimizes the costs of supply for the next day, and that also takes into account *bounds on the total number of hours that a small generator can be used during the whole year*.

The following information is available:

(i) The current quota (L, M, H) and corresponding unit prices in the contract with the power plants.
(ii) The contracts with the small generators, including restrictions on the total usage per year.
(iii) An accurate prediction of the electricity demand for the next day, and probabilistic information on demand during the remainder of the year.

In practice, the prediction of demand for the next day is accurate enough to be treated as deterministic information, denoted by d_t, $t \in T$. Hence, also this problem can be formulated as a two-stage integer recourse model:

$$\begin{aligned}
\min_{x,y,z} \quad & \sum_{t\in T}\left(\sum_{i=1}^{4} c^i x_t^i + \sum_{j\in J}\left(q^j y_t^j + r^j z_t^j\right)\right) + Q(y) \\
\text{s.t.} \quad & \sum_{i=1}^{4} x_t^i + \sum_{j\in J} b^j y_t^j \geq d_t, && t\in T \\
& x_t^1 \leq L,\ x_t^2 \leq M,\ x_t^3 \leq H,\ t\in T \\
& (y^j, z^j) \in C^j, && j\in J \\
& x_t^i \geq 0, && i = 1,\dots,4,\ t\in T \\
& y_t^j, z_t^j \in \{0,1\}, && j\in J,\ t\in T,
\end{aligned}$$

where the function Q reflects the expected recourse costs due to not meeting the restrictions on the yearly usage of the small generators. Below we propose a *simple recourse* structure to model these future costs.

Note that the first stage of the current model is exactly the same as the second-stage part of the previous model.

8.8.6 Future Demand for Supply from Small Generators

The simple recourse model that we have in mind assigns penalty costs to a surplus or shortage of yearly usage for each small generator. To model this, we need to translate the known distribution of future demand for electricity to a distribution of future demand for supply from the small generators.

Consider again a collection of representative days $R^1, \dots, R^N$, as defined before. For each R^i, with its typical demand patterns, we compute the optimal supply schedules giving in particular the optimal usage h^{ij} of each small generator SG^j, $j \in J$. On day τ of the year, let the random vector $\eta_\tau = (\eta_\tau^1, \dots, \eta_\tau^N)$ denote the remainder of the year, with η_τ^i the number of days of type i (obviously, $\sum_{i=1}^N \eta_\tau^i = 365 - \tau$). Given the probabilities of the realizations of η_τ, we can compute the distribution of the number of hours ξ_τ^j that supply from SG^j is needed in the remainder of the year, since

$$\xi_\tau^j = \sum_{i=1}^{N} h^{ij} \eta_\tau^i.$$

By this preprocessing step we can obtain the distribution of the random vector $\xi_\tau = (\xi_\tau^1, \dots, \xi_\tau^{|J|})$ for each day τ.

8.8.7 Modeling Expected Penalty Costs

Let L^j and U^j be the lower and upper bound on the total number of hours that SG^j can be used during the contract year. At day τ we know how much supply is taken from SG^j so far, giving current bounds

$$L_\tau^j := L^j - \sum_{s=1}^{\tau-1} u_s^j$$
$$U_\tau^j := U^j - \sum_{s=1}^{\tau-1} u_s^j,$$

where u_s^j is the number of hours that SG^j was used on day s. Thus, for given usage $u_\tau = (u_\tau^1, \dots, u_\tau^{|J|})$ and a realization ξ_τ of desired future supply by the small generators, the value function

$$\begin{aligned} v(u_\tau, \xi_\tau) := \min\ & q^+ y^+ + q^- y^- \\ \text{s.t.}\ & y^+ \geq \xi_\tau + u_\tau - U_\tau \\ & y^- \geq L_\tau - (\xi_\tau + u_\tau) \\ & y^+, y^- \in \mathbb{R}_+^{|J|} \end{aligned} \tag{8.4}$$

gives the minimal penalty costs for not meeting the yearly lower and upper bounds for each of the small generators. Since any surplus can be bought at unit costs c^4 from the power plants, we take all components of q^+ equal to c^4. On the other hand, if the total yearly supply from SG^j falls below the lower bound L^j, it seems reasonable to charge unit penalty costs equal to a fraction of $q^j + r^j$.

Due to the *simple recourse* structure, this second-stage problem is separable. Hence, the function v gives the total penalty costs as the sum of the penalty costs for each small generator individually:

$$v(u_\tau, \xi_\tau) = \sum_{j=1}^{|J|} v^j(u_\tau^j, \xi_\tau^j),$$

where v^j is the one-dimensional version of (8.4). Consequently, also the expected penalty costs function $Q(u_\tau) := \mathbb{E}_{\xi_\tau}[v(u_\tau, \xi_\tau)]$ is separable, with terms $Q^j(u_\tau^j) = \mathbb{E}_{\xi_\tau^j}\left[v^j(u_\tau^j, \xi_\tau^j)\right]$.

Using this penalty costs model may lead to the undesirable result that already early in the year the supply obtained from one or more small generators is close to one of its bounds, thus limiting the options for scheduling in the rest of the year. (Even if such a sequence of supply schedules has minimal total expected costs, it is undesirable in the sense that it does not correspond to current practice.) This effect can be avoided by means of the following refinement of the second-stage problem. The idea is to 'aim' total expected supply from each SG^j at the average value $(L^j_\tau + U^j_\tau)/2$, which is achieved by putting a small penalty on deviations from this value. In the refined model, penalty costs are given by

$$\begin{aligned} w(u_\tau, \xi_\tau) = \min\ & q^{1+}y^{1+} + q^{2+}y^{2+} + q^{1-}y^{1-} + q^{2-}y^{2-} \\ \text{s.t.}\ & y^{1+} + y^{2+} - y^{1-} - y^{2-} = \xi_\tau + u_\tau - (L_\tau + U_\tau)/2 \\ & y^{1+} \leq (U_\tau - L_\tau)/2 \\ & y^{1-} \leq (U_\tau - L_\tau)/2 \\ & y^{1+},\ y^{2+},\ y^{1-},\ y^{2-} \in \mathbb{R}^{|J|}_+, \end{aligned}$$

where q^{2+} and q^{2-} are equal to q^+ and q^- in the previous model, and the values of $q^{1+} < q^{2+}$ and $q^{1-} < q^{2-}$ are for example determined by numerical experiments (and are possibly decreasing with τ).

This refined model is called a *multiple simple recourse* model (see Sect. 3.1), and has the same separability properties as the simple recourse model.

8.8.8 *Optimal Supply Schedules*

Putting the ingredients discussed above together, we see that the problem of finding an optimal supply schedule for the next day can be modeled as a two-stage (multiple) simple recourse model, with mixed-binary first-stage variables and continuous second-stage variables.

Appendix A
Probability

In this appendix we give an incomplete description of elements of probability theory that are used in this book. Emphasis is on definitions in convexity and differentiability of so-called expected functions. Such functions arise everywhere in stochastic programming. The second part is the background of some results in Theorem 5.2.8, and the third part discusses several versions of probability distribution functions and probability density functions.

A.1 Expected Functions

Let ω be a random vector with support $\Omega \subset \mathbb{R}^r$ and cumulative distribution function (cdf) F. The expectation with respect to ω of a function $g(\omega, x)$ from $\mathbb{R}^r \times \mathbb{R}^n$ to $[-\infty, \infty]$ is given by

$$\mathbb{E}_\omega [g(\omega, x)] = \int_\Omega g(t, x)\, dF(t), \quad x \in \mathbb{R}^n, \tag{A.1}$$

where the right-hand side is a so-called Riemann-Stieltjes integral. Note that this is an r-fold integral. For example, if ω_i, $i = 1, \dots, r$, are independent random variables with support Ω_i and cdf F_i, then (A.1) can be written as

$$\begin{aligned}\mathbb{E}_\omega [g(\omega, x)] &= \int_{\Omega_1} \dots \int_{\Omega_r} g(t, x)\, dF_r(t_r) \dots dF_1(t_1) \\ &= \int_{-\infty}^{\infty} \dots \int_{-\infty}^{\infty} g(t, x)\, dF_r(t_r) \dots dF_1(t_1), \quad x \in \mathbb{R}^n.\end{aligned}$$

If we restrict ourselves to the case $r = 1$, then the meaning of (A.1) can be described as follows.

W. K. Klein Haneveld et al., *Stochastic Programming*, Graduate Texts in Operations Research, https://doi.org/10.1007/978-3-030-29219-5

(i) If ω has a probability density function (pdf) f, then

$$\frac{dF(t)}{dt} = f(t).$$

In (A.1) we substitute $dF(t) = f(t)\,dt$, giving

$$\mathbb{E}_\omega\,[g(\omega, x)] = \int_{-\infty}^{\infty} g(t, x) f(t)\,dt.$$

(ii) If ω follows a discrete distribution with $\Pr\left\{\omega = \omega^i\right\} = p^i,\ i \in I \subset \mathbb{Z}$, then (A.1) means

$$\mathbb{E}_\omega\,[g(\omega, x)] = \sum_{i \in I} p^i g(\omega^i, x).$$

Informally, this is justified by $p^i \simeq \Delta F(\omega^i)$.

(iii) The interpretation of (A.1) in case of a mixed distribution of ω is analogous to (i) and (ii).

Return to the general case in (A.1). We will call the function

$$\bar{g}(x) := \mathbb{E}_\omega\,[g(\omega, x)]\,, \quad x \in \mathbb{R}^n,$$

the *expected function* of g. It is not difficult to see, that this function is not well-defined in some special cases. For instance, if for some x we have $\Pr\{g(\omega, x) = -\infty\} > 0$ and $\Pr\{g(\omega, x) = +\infty\} > 0$ then the expected value of $g(\omega, x)$ does not exist, since $\infty - \infty$ is not defined. This difficulty can even occur if g is finite everywhere. Indeed, if $\mathbb{E}_\omega\,[g(\omega, x)]^+ = \infty$ and $\mathbb{E}_\omega\,[g(\omega, x)]^- = \infty$ then $\mathbb{E}_\omega\,[g(\omega, x)] \equiv \mathbb{E}_\omega\,[g(\omega, x)]^+ - \mathbb{E}_\omega\,[g(\omega, x)]^-$ is undefined. Therefore, when discussing expected functions we will implicitly assume that one of the three following cases is true for any $x \in \mathbb{R}^n$:

(a) $\mathbb{E}_\omega\,[g(\omega, x)]^+ < \infty$ and $\mathbb{E}_\omega\,[g(\omega, x)]^- < \infty$, so that $\bar{g}(x) \in \mathbb{R}$
(b) $\mathbb{E}_\omega\,[g(\omega, x)]^+ = \infty$ and $\mathbb{E}_\omega\,[g(\omega, x)]^- < \infty$, so that $\bar{g}(x) = +\infty$
(c) $\mathbb{E}_\omega\,[g(\omega, x)]^- < \infty$ and $\mathbb{E}_\omega\,[g(\omega, x)]^- = \infty$, so that $\bar{g}(x) = -\infty$

Then the expected function is a well-defined extended real function on $\mathbb{R}^n$.

We conclude this section with a few useful theorems for expected functions. The first one generalizes Corollary B.20.

Theorem A.1.1 *Let ω be a random vector with support $\Omega \subset \mathbb{R}^r$, and g an extended real function on $\mathbb{R}^r \times \mathbb{R}^n$, such that the expected function $\bar{g}(x) := \mathbb{E}_\omega\,[g(\omega, x)]$ is well-defined. Then $\bar{g}$ is a convex function if the function $g(t, \cdot)$ of $x \in \mathbb{R}^n$ is convex for all fixed $t \in \Omega$.*

Remark A.1.2 The statement remains true, if $g(t, \cdot)$ is convex for *almost* all t, that is, $\Pr\{t \in \Omega : g(t, \cdot) \text{ is not a convex function of } x\} = 0$.

The second theorem generalizes Theorem B.22. It contains the famous *Jensen's Inequality*. Notice that now convexity with respect to ω rather than with respect to x is assumed.

Theorem A.1.3 (Jensen's Inequality) *Let ω be a random vector with support $\Omega \subset \mathbb{R}^r$, and assume that its mean value $\mu = \mathbb{E}_\omega[\omega]$ is finite. If f is a proper convex function $\mathbb{R}^r \mapsto (-\infty, \infty]$ such that μ is in the relative interior of dom f, then*

$$\mathbb{E}_\omega[f(\omega)] \geq f(\mathbb{E}_\omega[\omega]) = f(\mu).$$

Proof The condition $\mu \in \operatorname{ri}\operatorname{dom} f$ implies that $\partial f(\mu) \neq \emptyset$. Hence, for $u \in \partial f(\mu)$ we have

$$f(t) \geq f(\mu) + \langle u, t - \mu\rangle \text{ for all } t \in \mathbb{R}^r.$$

Replacing t by ω, and taking expectations, we get the desired result, since

$$\mathbb{E}_\omega[\langle u, \omega - \mu\rangle] = \langle u, \mathbb{E}_\omega[\omega] - \mu\rangle = \langle u, 0\rangle = 0.$$

□

Corollary A.1.4 *Let ω be a random vector with support $\Omega \subset \mathbb{R}^r$ and finite mean value vector $\mu \in \mathbb{R}^r$, and g an extended real function on $\mathbb{R}^r \times \mathbb{R}^n$, such that the expected function $\bar{g}(x) := \mathbb{E}_\omega[g(\omega, x)]$ is well-defined. Then*

$$\bar{g}(x) \geq g(\mu, x), \quad x \in \mathbb{R}^n,$$

if for any fixed $x \in \mathbb{R}^n$, $g(t, x)$ is a convex function of t with μ in the relative interior of its domain.

Example A.1.5 Take $r = n = 1$. Then, for all $x \in \mathbb{R}$,

$$\mathbb{E}_\omega\left[(\omega - x)^2\right] \geq (\mu - x)^2$$
$$\mathbb{E}_\omega\left[(\omega - x)^+\right] \geq (\mu - x)^+$$
$$\mathbb{E}_\omega\left[(\omega - x)^-\right] \geq (\mu - x)^-$$

◇

The last theorem deals with derivatives of expected functions. It indicates, that if certain weak but incompletely specified conditions are valid, derivatives may be calculated by 'differentiation under the integral'. For simplicity, we state the result only for $n = 1$.

Theorem A.1.6 *Let ω be a random vector with support $\Omega \subset \mathbb{R}^r$, and g an extended function on $\mathbb{R}^r \times \mathbb{R}$, such that the expected value function $\bar{g}(x) := \mathbb{E}_\omega[g(\omega, x)]$ is* *well-defined and finite in the neighborhood of* $x_0 \in \mathbb{R}$. *If* g *is differentiable with respect to* x *at* $x = x_0$ *for almost all values of* ω, *i.e.* $\Pr\{\omega \in \Omega_0\} = 1$ *where*

$$\Omega_0 := \{t \in \Omega : \frac{\partial g}{\partial x}(t, x^0) \text{ exists as a finite number}\},$$

then, under some weak boundedness conditions it follows that also the expected function $\bar{g}$ *is differentiable at* x^0, *and*

$$\frac{d\bar{g}}{dx}(x^0) = \mathbb{E}_\omega\left[\frac{\partial g}{\partial x}(\omega, x^0)\right].$$

This statement remains true, if everywhere 'differentiable' is replaced by 'differentiable from the left'. Similar for 'differentiable from the right'.

Example A.1.7 Consider the function $g(t, x) = (t - x)^-$, $t \in \mathbb{R}$, $x \in \mathbb{R}$. For any fixed t, it is everywhere differentiable from the left with respect to x, with left derivatives

$$\frac{\partial g_-}{\partial x}(t, x) = \begin{cases} 0 \text{ if } x \le t, \\ 1 \text{ if } x > t. \end{cases}$$

Hence, also $\bar{g}(x) = \mathbb{E}_\omega[g(\omega, x)]$ is everywhere differentiable from the left, with left derivatives

$$\bar{g}'_-(x) = 0 \cdot \Pr\{x \le \omega\} + 1 \cdot \Pr\{x > \omega\} = \Pr\{\omega < x\}, \quad x \in \mathbb{R}.$$

Similarly for the right derivatives

$$\frac{\partial g_+}{\partial x}(t, x) = \begin{cases} 0 \text{ if } x < t, \\ 1 \text{ if } x \ge t, \end{cases}$$

so that

$$\bar{g}'_+(x) = 0 \cdot \Pr\{x < \omega\} + 1 \cdot \Pr\{x \ge \omega\} = \Pr\{\omega \le x\}, \quad x \in \mathbb{R}.$$

For the two-sided derivative we get

$$\frac{\partial g}{\partial x}(t, x) = \begin{cases} 0 & \text{if } x < t, \\ \text{undefined} & \text{if } x = t, \\ 1 & \text{if } x > t, \end{cases}$$

so that $\bar{g}$ is differentiable at x_0 *if* $\Pr\{\omega = x_0\} = 0$, with derivative

$$\bar{g}'(x_0) = \Pr\{\omega < x_0\} = \Pr\{\omega \le x_0\}$$

in that case. ◇

A.2 Limit Sets

Let $(\Omega, \mathcal{B}, P)$ be a probability space with events $S_1, S_2, \dots \in \mathcal{B}$.

Lemma A.2.1 *If $S_1 \subset S_2 \subset \dots$ then*

$$\lim_{n\to\infty} S_n := \bigcup_{k=1}^{\infty} S_k \in \mathcal{B}.$$

If $S_1 \supset S_2 \supset \dots$ then

$$\lim_{n\to\infty} S_n := \bigcap_{k=1}^{\infty} S_k \in \mathcal{B}.$$

In both cases

$$\Pr\left\{\lim_{n\to\infty} S_n\right\} = \lim_{n\to\infty} \Pr\{S_n\}.$$

□

Without monotonicity of the sets $S_1, S_2, \dots$ the following more general result holds.

Lemma A.2.2 *For a sequence $\{S_n\}$, define the* limit inferior

$$\begin{aligned}\underline{S} &:= \liminf_{n\to\infty} S_n = \bigcup_{n=1}^{\infty}\bigcap_{k=n}^{\infty} S_k \\ &= \{\omega \in \Omega : \omega \textit{ lies in all but finitely many } S_k\} \in \mathcal{B}\end{aligned}$$

and the limit superior

$$\begin{aligned}\overline{S} &:= \limsup_{n\to\infty} S_n = \bigcap_{n=1}^{\infty}\bigcup_{k=n}^{\infty} S_k \\ &= \{\omega \in \Omega : \omega \textit{ lies in infinitely many } S_k\} \in \mathcal{B}.\end{aligned}$$

Then

$$\Pr\{\underline{S}\} = \lim_{k\to\infty}\left(\Pr\left\{\bigcap_{n\ge k} S_n\right\}\right) \le \lim_{k\to\infty}\left(\inf_{n\ge k}\Pr\{S_n\}\right) = \liminf_{n\to\infty}\Pr\{S_n\}$$

$$\Pr\{\overline{S}\} = \lim_{k\to\infty}\left(\Pr\left\{\bigcup_{n\ge k} S_n\right\}\right) \ge \lim_{k\to\infty}\left(\sup_{n\ge k}\Pr\{S_n\}\right) = \limsup_{n\to\infty}\Pr\{S_n\}$$

Since $\underline{S} \subset \overline{S}$ *it holds* $\Pr\{\underline{S}\} \leq \Pr\{\overline{S}\}$, *in general with strict inequality. If equality holds, then*

$$\lim_{n\to\infty} S_n = \underline{S} = \overline{S}$$

is the limit of the sequence $\{S_n\}$. □

A.3 Probability Distribution and Density Functions

In this section we discuss the different versions of probability distribution functions and probability density functions that we use throughout this book.

Let ω be a one-dimensional random variable. Then, its (cumulative) probability distribution function F is defined as

$$F(x) = \Pr\{\omega \leq x\}, \quad x \in \mathbb{R}.$$

The distribution function F is non-decreasing and right-continuous with support contained in $[0, 1]$. We denote its left-continuous version by $\tilde{F}$, where

$$\tilde{F}(x) = \Pr\{\omega < x\}, \quad x \in \mathbb{R}.$$

Obviously, $F(x) = \tilde{F}(x)$ for all $x \in \mathbb{R}$ if ω is continuously distributed. In this case, ω has a probability density function f, satisfying

$$F(x) = \int_{-\infty}^{x} f(t)dt, \quad x \in \mathbb{R}. \tag{A.2}$$

Interestingly, f is not uniquely defined. As can be observed from (A.2), we can change f at finitely many (or even countably many) points, to obtain an alternative density function $\hat{f}$ with

$$\int_{-\infty}^{x} f(t)dt = \int_{-\infty}^{x} \hat{f}(t)dt, \quad x \in \mathbb{R}.$$

In probability theory, this technicality is not important. All versions of the density functions are equivalent because they determine the same distribution. On the other hand, from other points of view (e.g. continuity) different versions behave differently. That is also the case in Chap. 4. For that reason, we restrict the attention to 'proper' continuous distributions for which a unique 'regular version' of the density exist. That is

- We call a distribution *proper* if the distribution has a pdf that is a continuous function, except in at most countably infinite discontinuity points, in such a way that each bounded interval only contains finitely many discontinuity points.

- We call a density function *regular* if it is right-continuous and corresponds to a proper distribution.

For applications the restriction to proper distributions and regular probability density functions is harmless. That is why we only consider such distributions and pdfs in this book. We use $\mathcal{R}$ to denote the class of regular density functions.

An advantage of this definition is that among all possible equivalent density functions, regular density functions have the smallest *total variation*.

Definition A.3.1 Let f be a real function on a non-empty subset I of $\mathbb{R}$. Then the *total increase* $\Delta^+ f(I)$*, total decrease* $\Delta^- f(I)$*, and total variation* $|\Delta| f(I)$ *of* f *on* I are defined as

$$\Delta^+ f(I) = \sup_U \sum_{i=1}^{n} \Big(f(u_i) - f(u_{i-1}) \Big)^+ ,$$

$$\Delta^- f(I) = \sup_U \sum_{i=1}^{n} \Big(f(u_i) - f(u_{i-1}) \Big)^- ,$$

$$|\Delta| f(I) = \sup_U \sum_{i=1}^{n} \Big| f(u_i) - f(u_{i-1}) \Big| ,$$

respectively, where the supremum is taken over all finite subsets $U = \{u_0, u_1, \ldots, u_n\} \subset I$ such that $u_0 < u_1 < \cdots < u_n$.

For convenience we use the following shorthand notations for the case $I = \mathbb{R}$:

$$\begin{aligned}
\Delta^+ f &= \Delta^+ f((-\infty, \infty)) \\
\Delta^- f &= \Delta^- f((-\infty, \infty)) \\
|\Delta| f &= |\Delta| f((-\infty, \infty)).
\end{aligned}$$

For all $I \subset (-\infty, \infty)$ it holds $|\Delta| f(I) = \Delta^+ f(I) + \Delta^- f(I)$.

Definition A.3.2 A real function f on $\mathbb{R}$ is *of bounded variation* if $|\Delta| f < +\infty$.

Appendix B
Elementary Convex Analysis

This appendix contains an overview of relevant results from Convex Analysis. Almost all results are quoted from Rockafellar [35], where the reader is referred to for proofs.

As we will see below, there is an intimate relation between convex sets and convex functions. Indeed, we will define convex functions in terms of convex sets. The main reason to do so is that many results on convex functions follow easily from corresponding results on convex sets.

Definition B.1 A *subset* C of $\mathbb{R}^n$ is *convex* if $(1-\lambda)x + \lambda y \in C$ whenever $x \in C$, $y \in C$ and $0 < \lambda < 1$.

Definition B.2 A vector sum $\lambda_1 x_1 + \cdots + \lambda_m x_m$ is a *convex combination* of $x_1, \ldots, x_m$ if the coefficients λ_i are all non-negative and $\lambda_1 + \cdots + \lambda_m = 1$.

Theorem B.3 *A subset of $\mathbb{R}^n$ is convex if and only if it contains all convex combinations of its elements.*

In many situations (e.g. in Stochastic Programming) $\lambda_1, \ldots, \lambda_m$ can be interpreted as probabilities. For instance, if ω is a random vector in $\mathbb{R}^n$ that takes the values $x_i \in \mathbb{R}^n$ with probability λ_i, $i = 1, \ldots, m$, (with $\lambda_i > 0,\ \lambda_1 + \cdots + \lambda_m = 1$), then the convex combination $\lambda_1 x_1 + \cdots + \lambda_m x_m$ is precisely the mean value $\mathbb{E}[\omega]$ of ω.

Before we define what a convex function is, we will indicate why we do not wish to restrict to finite functions.

Our interest in convex functions is motivated by *constrained* infimization problems. For theoretical considerations it is convenient to incorporate the constraints into the objective function by (re-)defining the function to be equal to $+\infty$ in infeasible points. That is, let $\hat{f}$ be a function which we wish to infimize over a constraint set C. Then the optimal solution(s) and optimal value do not change if we

W. K. Klein Haneveld et al., *Stochastic Programming*, Graduate Texts in Operations Research, https://doi.org/10.1007/978-3-030-29219-5

consider the *unconstrained* infimization of

$$f(x) = \begin{cases} \hat{f}(x), & x \in C; \\ \infty, & \text{otherwise.} \end{cases} \tag{B.1}$$

Definition B.4 A map f defined on (a subset of) $\mathbb{R}^n$ is an *extended real function* if its values are real or $-\infty$ or ∞.

We adopt the following rules for calculations involving $+\infty$ and $-\infty$:

$$\begin{aligned}
&-\infty < \alpha < \infty && \forall \alpha \in \mathbb{R} \\
&\alpha + \infty = \infty + \alpha = \infty, && -\infty < \alpha \leq \infty \\
&\alpha - \infty = -\infty + \alpha = -\infty, && -\infty \leq \alpha < \infty \\
&\alpha \cdot \infty = \infty \cdot \alpha = \infty, \quad \alpha \cdot (-\infty) = (-\infty) \cdot \alpha = -\infty, && 0 < \alpha \leq \infty \\
&\alpha \cdot \infty = \infty \cdot \alpha = -\infty, \quad \alpha \cdot (-\infty) = (-\infty) \cdot \alpha = \infty, && -\infty \leq \alpha < 0 \\
&0 \cdot \infty = \infty \cdot 0 = 0 = 0 \cdot (-\infty) = (-\infty) \cdot 0, \quad -(-\infty) = \infty && \\
&\inf \emptyset = +\infty, \qquad \sup \emptyset = -\infty && \\
&\sum_{i \in \emptyset} \alpha_i = 0, \qquad \prod_{i \in \emptyset} \alpha_i = 1 &&
\end{aligned}$$

The combinations $\infty - \infty$ and $-\infty + \infty$ are undefined and are avoided (see e.g. Definition B.8).

Often, extended real functions come up in a natural way as the optimal value function of optimization problems.

Example B.5 Consider a linear programming problem

$$\min_{y \in \mathbb{R}^p} \{qy : \; Wy = z, \; y \geq 0\},$$

where q is a p-dimensional row vector, W is an $m \times p$ matrix and $z \in \mathbb{R}^m$. The optimal value function of this LP problem is defined as the following extended real function on $\mathbb{R}^m$

$$v(z) = \inf_{y \in \mathbb{R}^p} \{qy : \; Wy = z, \; y \geq 0\}, \quad z \in \mathbb{R}^m.$$

Basically, there are three cases:

(i) If there are no vectors y satisfying all constraints (i.e., the problem is infeasible) then $v(z) = +\infty$ for such values of z.
(ii) If there exist feasible solutions, such that arbitrary low values of the objective function can be obtained (i.e., the problem has an 'unbounded solution') then $v(z) = -\infty$ for such values of z.

(iii) If the problem is feasible, and the objective function is bounded from below on the feasible set, then $v(z) \in \mathbb{R}$ for such a z. In this case (and only in this case!) the infimum is attained (i.e., there exists at least one optimal solution, and the 'infimum' can be replaced by 'minimum').

$\diamond$

From now on we will call 'extended real functions' just 'functions'. In particular, we are interested in convex functions.

The following definition identifies a relation between functions and certain sets. Our definition of convex functions is based on this relation.

Definition B.6 Let f be a function on $\mathbb{R}^n$. The set

$$\{(x, \nu) \in \mathbb{R}^n \times \mathbb{R} : \nu \geq f(x)\}$$

is the *epigraph* of f and is denoted by epi f.

Note that epi f is a set in $\mathbb{R}^{n+1}$ not containing infinite points.

Now we are ready to define convex functions.

Definition B.7 A *function* f is *convex* if the set epi f is convex.

This definition is also valid if f takes on values $-\infty$ and/or ∞. We have already seen that the value $+\infty$ plays an important role. However, in practice exclusion of the value $-\infty$ is often quite practical.

Definition B.8 A convex function f is *proper* if $f(x) < +\infty$ for at least one x and $f(x) > -\infty$ for every x.

For example, if $\hat{f}$ is convex and finite on a nonempty convex set C, then the function f as defined by (B.1) is a proper convex function.

Another useful concept in this context is

Definition B.9 The *effective domain* of a convex function f, denoted by dom f, is the projection on $\mathbb{R}^n$ of the epigraph of f:

$$\text{dom } f = \{x : \exists \nu, (x, \nu) \in \text{epi } f\} = \{x : f(x) < +\infty\}.$$

For example, in (B.1) we have dom $f = C$. Of course, convexity of f is equivalent to that of the restriction of f to dom f.

If we exclude the value $-\infty$, Definition B.7 is equivalent to the following useful result which is often used as the definition of a convex function.

Theorem B.10 *Let f be a function from $C \subset \mathbb{R}^n$ to $(-\infty, \infty]$. Then f is convex on C if and only if C is a convex set and*

$$f((1-\lambda)x + \lambda y) \leq (1-\lambda)f(x) + \lambda f(y), \qquad 0 < \lambda < 1,$$

for every x and y in C.

As examples of results on convex sets that imply results on convex functions, we present the following.

Theorem B.11 *The intersection of an arbitrary collection of convex sets is convex.*

This is very easy to prove. However, it implies the following maybe not so obvious result.

Corollary B.12 *The pointwise supremum of an arbitrary collection of convex functions is convex.*

Similarly, Theorem B.13 implies Corollary B.14.

Theorem B.13 *The orthogonal projection of a convex set C on a subspace L is convex set.*

Corollary B.14 *Let f be a convex function on $\mathbb{R}^n \times \mathbb{R}^m$. Then*

$$h(x) = \inf_{y \in \mathbb{R}^m} f(x, y), \qquad x \in \mathbb{R}^n,$$

is a convex function.

For example, the optimal value function of the LP problem as defined in Example B.5 is convex.

Together, Theorems B.11 and B.13 imply

Theorem B.15 *For any convex function f and $\alpha \in [-\infty, \infty]$ the level sets $\{x : f(x) < \alpha\}$ and $\{x : f(x) \leq \alpha\}$ are convex.*

As a last example, we discuss the results of a linear transformation. Given a linear transformation A from $\mathbb{R}^n$ to $\mathbb{R}^m$, we define

$$AC = \{Ax : x \in C\}, \qquad C \subset \mathbb{R}^n.$$

Theorem B.16 *Let A be a linear transformation from $\mathbb{R}^n$ to $\mathbb{R}^m$. Then AC is a convex set in $\mathbb{R}^m$ for every convex set C in $\mathbb{R}^n$.*

Corollary B.17 *Let A be a linear transformation from $\mathbb{R}^n$ to $\mathbb{R}^m$. Then, for each convex function f on $\mathbb{R}^m$, the function fA defined by*

$$(fA)(x) = f(Ax)$$

is convex on $\mathbb{R}^n$.

Next we consider the (weighted) sum of proper convex functions.

Theorem B.18 *If f_1 and f_2 are proper convex functions on $\mathbb{R}^n$, then $f_1 + f_2$ is convex.*

Corollary B.19 *A linear combination* $\lambda_1 f_1 + \cdots + \lambda_m f_m$ *of proper convex functions with non-negative coefficients is convex.*

Due to the fact that we restrict ourselves to proper functions, summations involving both $+\infty$ and $-\infty$ are avoided. Therefore, the resulting functions are always well defined. However, they may not be proper.

On interpreting the (normalized) coefficients as probabilities we obtain for finite discrete distributions the following result.

Corollary B.20 *Let* ω *be a random vector with mass points* ω_i*, and* $\Pr\{\omega = \omega_i\} = p_i$, $i = 1, \ldots, m$. *If* $f(\omega, x)$ *is a proper convex function in* x *for every* ω_i*, then*

$$\mathbb{E}_\omega [f(\omega, x)] = \sum_{i=1}^{m} p_i f(\omega_i, x)$$

is convex in x.

Exercise B.21 Show that Corollary B.20 also holds if ω has a density. ◁

A well known result that also has a clear interpretation in terms of probabilities is

Theorem B.22 *Let* f *be a function from* $\mathbb{R}^n$ *to* $(-\infty, \infty]$. *Then* f *is convex if and only if*

$$f(\lambda_1 x_1 + \cdots + \lambda_m x_m) \leq \lambda_1 f(x_1) + \cdots + \lambda_m f(x_m)$$

whenever $\lambda_1 \geq 0, \ldots, \lambda_m \geq 0$, $\lambda_1 + \cdots + \lambda_m = 1$.

The following two results are of use to check whether a certain given function is convex or not.

Theorem B.23 *Let* f *be a twice continuously differentiable real-valued function on an open convex set* C *in* $\mathbb{R}^n$. *Then* f *is convex on* C *if and only if the Hessian matrix*

$$Q_x = (q_{ij}(x)), \qquad q_{ij}(x) = \frac{\partial^2 f}{\partial \xi_i \partial \xi_j}(\xi_1, \ldots, \xi_n)$$

is positive semi-definite for every $x = (\xi_1, \xi_2 \ldots, \xi_n)' \in C$.

Formulation of the one-dimensional version of Theorem B.23 is left as an exercise.

Theorem B.24 *Let* f *be a convex function from* $\mathbb{R}^n$ *to* $(-\infty, \infty]$*, and let* φ *be a convex function from* $\mathbb{R}$ *to* $(-\infty, \infty]$ *which is non-decreasing. Then* $h(x) = \varphi(f(x))$ *is convex on* $\mathbb{R}^n$ *(where one sets* $\varphi(\infty) = \infty$*).*

For example, it follows that $e^{f(x)}$ is a proper convex function on $\mathbb{R}^n$ if f is. Also, if g is concave (see Definition B.34 below), then $h(x) = 1/g(x)$ is convex on $C = \{x : g(x) > 0\}$.

Now we will discuss an important concept that generalizes differentiability, at least for convex functions.

Definition B.25 A vector $x^* \in \mathbb{R}^n$ is a *subgradient* of a function f at a point $x \in \mathbb{R}^n$ if

$$f(z) \geq f(x) + \langle x^*, z - x \rangle \qquad \forall z \in \mathbb{R}^n$$

where $\langle \cdot, \cdot \rangle$ denotes the inner product in $\mathbb{R}^n$.

An interpretation of this so-called *subgradient inequality* is that the affine function $h(z) = f(x) + \langle x^*, z - x \rangle$ is a non-vertical supporting hyperplane to the set epi f at the point $(x, f(x))$, at least if $f(x)$ is finite.

The set of all subgradients of f at x is called the *subdifferential of f at x* and is denoted by $\partial f(x)$. If $\partial f(x)$ is not empty then f is said to be *subdifferentiable* at x.

The subdifferentials of f at points x where $f(x)$ is not finite are of little interest. Indeed,

$$\begin{aligned} &\text{if } f(x) = -\infty \text{ then} && \partial f(x) = \mathbb{R}^n \\ &\text{if } f(x) = +\infty \text{ then} && \begin{cases} \partial f(x) = \mathbb{R}^n \text{ if } f(z) = +\infty \quad \forall\, z \\ \partial f(x) = \emptyset \quad \text{if } \exists\, z,\ f(z) < +\infty \end{cases} \end{aligned}$$

In particular, if f is a *proper* convex function, then $\partial f(x) = \emptyset$ if $x \notin \operatorname{dom} f$. On the other hand we have for these functions the following positive result.

Theorem B.26 *If f is a proper convex function on $\mathbb{R}^n$, then $\partial f(x)$ is a nonempty closed convex set for all $x \in \operatorname{dom} f$, except maybe at some boundary points of* $\operatorname{dom} f$.

Remark B.27 A more precise statement is, that a proper convex function is subdifferentiable in all points of the so-called relative interior of its effective domain. If dom f is full-dimensional, the relative interior is just the interior; if not, the interior is empty but the relative interior is not. For instance, the set $C = \{x \in \mathbb{R}^2 : x_1 \in [0, 1], x_2 = 1\}$ has an empty interior, since it does not contain 2-dimensional balls, but its relative interior ri C is simply $\{x \in \mathbb{R}^2 : x_1 \in (0, 1), x_2 = 1\}$.

Example B.28 Define $f : \mathbb{R} \to (-\infty, \infty]$ by

$$f(x) = \begin{cases} |x|, \ x \in [-1, 1), \\ 2, \ \ x = 1, \\ \infty, \text{ otherwise.} \end{cases}$$

Then f is a proper convex function on $\mathbb{R}$, with dom $f = [-1, 1]$. It is easily verified that

$$\partial f(x) = \begin{cases} (-\infty, -1], & x = -1, \\ \{-1\}, & -1 < x < 0, \\ [-1, 1], & x = 0, \\ \{1\}, & 0 < x < 1, \\ \emptyset, & x = 1 \text{ (or } x < -1 \text{ or } x > 1). \end{cases}$$

Note that f is subdifferentiable at the left boundary point of dom f but not at the right boundary point. ◇

Example B.29 Define $f : \mathbb{R} \to (-\infty, \infty]$ by

$$f(x) = \begin{cases} \infty, & x < 0, \\ -\sqrt{x}, & x \geq 0. \end{cases}$$

Then f is a proper convex function on $\mathbb{R}$, with dom $f = [0, \infty)$. It is easily verified that

$$\partial f(x) = \begin{cases} \emptyset, & x \leq 0, \\ \left\{\frac{-1}{2\sqrt{x}}\right\}, & x > 0. \end{cases}$$

◇

The examples indicate an intimate relation between the subdifferential of f at x and its derivative (if it exists). Indeed it holds, that $\partial f(x)$ consists of a *single* vector x^* if and only if the convex function f is finite in the neighborhood of x, is differentiable at x and has x^* as its gradient $\nabla f(x) := \left(\frac{\partial f}{\partial \xi_1}(x), \ldots, \frac{\partial f}{\partial \xi_n}(x)\right)'$ with $x = (\xi_1, \xi_2, \ldots, \xi_n)' \in \mathbb{R}^n$. For one-dimensional proper convex functions, there is a straightforward relation between the subdifferential, and the left and right derivative functions of f, denoted by f'_- and f'_+ respectively.

Theorem B.30 *Let f be a proper convex function on $\mathbb{R}$. Then*

(i) $f'_-(x)$ *and* $f'_+(x)$ *are well-defined extended real numbers for all* $x \in$ dom f;
(ii) $f'_-(x)$ *and* $f'_+(x)$ *are non-decreasing functions on* dom f;
(iii) $f'_-(x) \leq f'_+(x)$ *for all* $x \in$ dom f;
(iv) $\partial f(x) = [f'_-(x), f'_+(x)] \cap \mathbb{R}$ *for all* $x \in$ dom f.

Notice that left and right derivatives may have infinite values, whereas a subgradient is finite, by definition. For instance, in Example B.29 we have that $f'_-(0) = f'_+(0) = -\infty$ and $\partial f(0) = \emptyset$. Similarly, in Example B.28 it holds that $f'_-(1) = f'_+(1) = +\infty$ and $\partial f(1) = \emptyset$, whereas $f'_-(-1) = -\infty$, $f'_+(-1) = -1$ and $\partial f(-1) = (-\infty, -1]$.

In case a function g is not convex, one may be interested in the following convex approximation of g.

Definition B.31 The *convex hull of a function* g, denoted by $\operatorname{conv} g$, is the pointwise greatest convex function majorized by g.

As suggested by its name, the convex hull of g can be characterized via the convex hull of sets.

Definition B.32 The *convex hull of* $S \subset \mathbb{R}^n$ is the intersection of all convex sets containing S.

Let $\operatorname{epi} g$ denote the epigraph of the function $g : \mathbb{R}^n \mapsto [-\infty, \infty]$. Then

$$\operatorname{conv} g(x) = \inf\{s : (x, s) \in \operatorname{conv}(\operatorname{epi} g)\},$$

where $\operatorname{conv} S$ denotes the convex hull of a set S.

The following properties of the convex hull make it very attractive in the context of minimization:

$$\inf_{x \in \mathbb{R}^n} g(x) = \inf_{x \in \mathbb{R}^n} \operatorname{conv} g(x),$$

and

$$\operatorname*{argmin}_{x \in \mathbb{R}^n} g(x) \subset \operatorname*{argmin}_{x \in \mathbb{R}^n} \operatorname{conv} g(x).$$

Notice that these properties have to do with *unconstrained* minimization of g over $\mathbb{R}^n$. If one is interested in infimization of g over a non-empty closed convex subset C of $\mathbb{R}^n$, one should replace g by $\bar{g}$, given by

$$\bar{g}(x) = \begin{cases} g(x), & \text{if } x \in C; \\ +\infty, & \text{if } x \notin C. \end{cases}$$

Obviously, *unconstrained* minimization of $\bar{g}$, or rather of $\operatorname{conv} \bar{g}$, is equivalent to the original constrained minimization.

We conclude this section by definitions of several related concepts. Each of them is used somewhere in the book. Here, we do not elaborate on them.

Definition B.33 A function f is *quasi-convex* if all lower level sets $\{x : f(x) \leq \alpha\}$, $\alpha \in \mathbb{R}$, are convex.

Clearly, if f is convex then f is quasi-convex. The reverse statement is not true, however.

Definition B.34 A function f is *concave* if $-f$ is convex.

Definition B.35 A nonnegative function f is *log-concave* if $\log(f)$ is concave. (Here $\log 0 := -\infty$)

A log-concave function f is quasi-concave (i.e. $-f$ is quasi-convex). Verify that the density function of the normal distribution is log-concave.

Definition B.36 A function f is *lower semi-continuous* (lsc) if

$$\liminf_{y \to x} f(y) \geq f(x) \quad \forall x \in \mathbb{R}^n.$$

Definition B.37 A function f is *upper semi-continuous* (usc) if

$$\limsup_{y \to x} f(y) \leq f(x) \quad \forall x \in \mathbb{R}^n.$$

Lemma B.38

(i) *f is lsc if and only if $-f$ is usc;*
(ii) *f is lsc if and only if all lower level sets $\{x : f(x) \leq \alpha\}$, $\alpha \in \mathbb{R}$, are closed;*
(iii) *f is lsc if and only if* epi f *is a closed set;*
(iv) *f is continuous if and only if f is both lsc and usc.*

Example B.39 Let

$$f(x) = \begin{cases} 1, \ x \in (0, 1); \\ c, \ x = 0 \text{ or } x = 1; \\ 0, \ \text{otherwise}. \end{cases}$$

If $c \leq 0$ then f is lsc; if $c \geq 1$ then f is usc; otherwise, f is neither lsc nor usc.

For instance, if $c < 0$ then the lower level sets are given by

$$\{x \in \mathbb{R} : f(x) \leq \alpha\} = \begin{cases} \emptyset, & \alpha < c; \\ \{0\} \cup \{1\}, & c \leq \alpha < 0; \\ (-\infty, 0] \cup [1, \infty), & 0 \leq \alpha < 1; \\ (-\infty, \infty), & \alpha \geq 1, \end{cases}$$

and all of them are closed, indeed. ◇

Definition B.40 A function f that is finite on a set C is *Lipschitz continuous* on C if there exists a constant L such that

$$|f(x) - f(y)| \leq L\|x - y\|$$

for all $x \in C$ and $y \in C$. Its *Lipschitz constant* is

$$\sup\left\{\frac{|f(x) - f(y)|}{\|x - y\|} : x \in C, \ y \in C, \ x \neq y\right\}.$$

For example, the function $f(x) = x^2$ is Lipschitz continuous on any bounded interval. However, it is not Lipschitz on $\mathbb{R}$.

Definition B.41 A real function on $\mathbb{R}$ is *semi-periodic* with period p and slope s if it is the sum of an affine function with slope s and a function that is periodic with period p.

Appendix C
Deterministic LP

Consider the function v,

$$v(z) = \inf_{y \in \mathbb{R}^p} \{qy : Wy = z,\ y \geq 0\}, \quad z \in \mathbb{R}^m,$$

the optimal value function of an LP problem in canonical form, as a function of the right-hand side vector $z \in \mathbb{R}^m$. Let us denote this LP problem by $P(z)$. The corresponding dual problem $D(z)$ reads

$$v^d(z) = \sup_{\lambda \in \mathbb{R}^m} \{\lambda z : \lambda W \leq q\}, \quad z \in \mathbb{R}^m,$$

where the vector λ of dual variables is denoted as a row vector in $\mathbb{R}^m$, and the optimal value of the dual problem, as a function of $z \in \mathbb{R}^m$, is denoted as $v^d(z)$.

Here, as in most parts of this book, transposition symbols are omitted. So, q and λ are implicitly assumed to be row vectors, whereas y and z are column vectors. So qy and λz are just inner products. Similarly, gradients are interpreted as row vectors.

Remark C.1 It is possible that optimal values are infinite:

- if $P(z)$ is infeasible, then $v(z) = +\infty$
- if $P(z)$ has arbitrary small values of the objective function on the feasible set, then $v(z) = -\infty$
- if $D(z)$ is infeasible, then $v^d(z) = -\infty$
- if $D(z)$ has arbitrary large values of the objective function on the feasible set, then $v^d(z) = +\infty$

In any such case, the infimum (c.q. supremum) is not attained.

The following proposition summarizes the duality theory of deterministic LP (part A), and gives a formula for the gradient of v in case it exists (part B).

W. K. Klein Haneveld et al., *Stochastic Programming*, Graduate Texts in Operations Research, https://doi.org/10.1007/978-3-030-29219-5

Proposition C.2

(A) For any fixed $z \in \mathbb{R}^m$ *we have*

$$-\infty \leq v^d(z) \leq v(z) \leq \infty \qquad \textit{(weak duality)}.$$

In fact, it holds that

$$-\infty \leq v^d(z) = v(z) \leq \infty$$

unless both $P(z)$ and $D(z)$ are infeasible (then $-\infty = v^d(z) < v(z) = \infty$). In particular, if both $P(z)$ and $D(z)$ are feasible then

$$-\infty < v^d(z) = v(z) < \infty$$

and the infimum (supremum, respectively) is attained (= min, max *respectively). That is, both problems have optimal solutions $y^\star$ ($\lambda^\star$, respectively), that are characterized by feasibility ($Wy^\star = z,\ y^\star \geq 0,\ \lambda^\star W \leq q$) and* complementary slackness

$$\sum_{i=1}^{m} \underbrace{\left(q - \lambda^\star W\right)_i}_{\geq 0} \underbrace{y_i^\star}_{\geq 0} = 0. \tag{C.1}$$

At least one of each pair $(q - \lambda^\star W)_i$ and $y_i^\star$ must be zero.

(B) For all $z \in \mathbb{R}^m$ together *we have in addition:*
The optimal value functions v and v^d are convex polyhedral functions (finite or not). They are finite all over $\mathbb{R}^m$ (and equal therefore) if and only if the following conditions hold:

(i) W is a complete recourse matrix (i.e. the primal problem $P(z)$ has feasible solutions for all $z \in \mathbb{R}^m$)
(ii) The recourse structure (Y, q, W), with $Y = \mathbb{R}^m_+$, is not extremely inexpensive (i.e. $\exists \lambda \in \mathbb{R}^m,\ q \geq \lambda W$; $D(z)$ has feasible solutions for all $z \in \mathbb{R}^m$).

In particular, if $z \in \mathbb{R}^m$ is such that the dual problem $D(z)$ has a unique *optimal solution (say $\lambda^\star$), then v is differentiable at z, and*

$$\nabla v(z) = \lambda^\star, \quad \text{i.e.}\ \frac{\partial v}{\partial z_i}(z) = \lambda_i^\star,\ i = 1, \ldots, m.$$

Proof Proof of part A is left as an exercise (see Assignment R2).

Under the assumptions (i) and (ii) in the proposition, it is not difficult to derive an explicit compact representation for the optimal value function v. Indeed, then the set

$$\Lambda := \{\lambda \in \mathbb{R}^m : \lambda W \leq q\}$$

must be a nonempty *bounded* (why?) polyhedral set in $\mathbb{R}^p$. Therefore it is spanned by its extreme points λ^k, $k = 1, \ldots, K$. Since $v(z) = v^d(z)$ for all $z \in \mathbb{R}^m$ in this case, we have

$$v(z) = \max_{k=1,\ldots,K} \lambda^k z, \quad z \in \mathbb{R}^m. \tag{C.2}$$

Therefore, v is the *pointwise maximum of a finite set of linear functions of* z. Moreover, the extreme point λ^j is optimal for the dual problem $D(z)$ if and only if

$$\lambda^j z \geq \lambda^k z \text{ for all } k = 1, \ldots, K.$$

In particular, λ^j is the unique optimal solution of $D(z)$ if z is such that

$$\lambda^j z > \lambda^k z \text{ for all } k \in \{1, \ldots, K\} \setminus \{j\}.$$

Then λ^j is also optimal for $D(z)$ if z is slightly perturbed, so that $\nabla v(z) = \lambda^j$ in that case. □

In the next proposition we present a characterization of the subdifferential $\partial v(z)$, that generalizes the last part of Proposition C.2.

Proposition C.3 *Suppose that the recourse structure* (q, W) *is not extremely inexpensive. Then*

$$-\infty < v(z) = v^d(z) \leq +\infty \quad \textit{for all } z \in \mathbb{R}^m, \tag{C.3}$$

so that v $(= v^d)$ *is a proper convex extended real function on* $\mathbb{R}^m$. *Then the subdifferential* $\partial v(z)$ *is nonempty for all* $z \in \operatorname{dom} v$, *and it is characterized as*

$$\partial v(z) = \Lambda^*(z) \quad \textit{for all } z \in \operatorname{dom} v, \tag{C.4}$$

where $\Lambda^*(z)$ *denotes the set of all optimal solutions of the problem* $D(z)$, *that is, the set of all shadow prizes for the equality constraints in* $P(z)$.

Proof Since (q, W) is assumed to be not extremely inexpensive, the problem $D(z)$ is feasible for all $z \in \mathbb{R}^m$, so that $v^d(z) > -\infty$ for all z. Consequently, (C.3) follows from the first part of Proposition C.2. The same assumption implies that $v(0) = 0$.

Exercise C.4 Why? Hint: Since $y = 0$ is feasible in $P(0)$, we have $v(0) \leq 0$. In fact, there are only two possible values for $v(0)$: it is either 0 or $-\infty$, since the constraints in $P(0)$ have right-hand sides equal to 0. But $v(0) = v^d(0) > -\infty$. ◁

As a consequence, $0 \in \operatorname{dom} v$ so that v is proper (and convex, see Proposition C.2 (B)).

Now consider any fixed $z^0 \in \operatorname{dom} v$. Then $v(z^0) = v^d(z^0)$ is finite, and from duality theory of LP we conclude that $\Lambda^*(z^0) \neq \emptyset$. It remains to show, that (C.4) holds. This is done in two parts.

(a) *Suppose that $\lambda^0 \in \Lambda^*(z^0)$. Then we will show $\lambda^0 \in \partial v(z^0)$.*

Since $\lambda^0 \in \Lambda^*(z^0)$ we know that $v(z^0) = \lambda^0 z^0$ and $\lambda^0 W \leq q$. Moreover, λ^0 is feasible for $D(z)$ for all $z \in \mathbb{R}^m$ so that $v(z) \geq \lambda^0 z$ for all $z \in \mathbb{R}^m$. Therefore,

$$v(z) \geq v(z^0) + \lambda^0(z - z^0) \quad \text{for all } z \in \mathbb{R}^m,$$

implying $\lambda^0 \in \partial v(z^0)$.

(b) *Suppose that $\lambda^0 \in \partial v(z^0)$. Then we will show $\lambda^0 \in \Lambda^*(z^0)$.*

Since $\lambda^0 \in \partial v(z^0)$, we have

$$v(z) \geq v(z^0) + \lambda^0(z - z^0) \quad \text{for all } z \in \mathbb{R}^m.$$

By rearranging terms this means

$$\min_{z \in \mathbb{R}^m} \left(v(z) - \lambda^0 z \right) = v(z^0) - \lambda^0 z^0. \tag{C.5}$$

The right-hand side is the finite optimal value of the left-hand side optimization problem. This problem can be written as a linear programming problem in (y, z):

$$\min_{\substack{y \in \mathbb{R}^p \\ z \in \mathbb{R}^m}} \left\{ qy - \lambda^0 z : Wy - z = 0,\ y \geq 0 \right\}.$$

Since the optimal value is finite, it is equal to the optimal value of the corresponding dual problem

$$\max_{\lambda \in \mathbb{R}^m} \left\{ \lambda \cdot 0 : \begin{array}{ll} \lambda W & \leq q \\ \lambda(-I) & = -\lambda^0 \end{array} \right\}.$$

Hence, it is equal to zero, and $q \geq \lambda^0 W$. Therefore, we did show that (C.5) implies

$$0 = v(z^0) - \lambda^0 z^0, \quad q \geq \lambda^0 W,$$

and this is nothing other than $\lambda^0 \in \Lambda^*(z^0)$. □

Appendix D
Maximum Calculus

In the context of various penalty models the following notations are useful. In each case $x \in \mathbb{R}$. (If the argument is more-dimensional then all relations hold too, if they are interpreted component-wise.)

$$
\begin{aligned}
(x)^+ &:= \max\{0, x\} \\
(x)^- &:= \max\{0, -x\} = -\min\{0, x\}
\end{aligned}
$$

The following relations are easily verified:

$$
\begin{aligned}
(-x)^- &= (x)^+ \\
(cx)^+ &= c(x)^+, \qquad c \in \mathbb{R}_+ \\
(x)^+ + (x)^- &= |x| \\
(x)^+ - (x)^- &= x \\
(x)^+ &= \big(|x| + x\big)/2 \\
(x)^- &= \big(|x| - x\big)/2 \\
(x)^+.(x)^- &= 0 \\
\max\{(x)^+, (x)^-\} &= |x| \\
\min\{(x)^+, (x)^-\} &= 0
\end{aligned}
$$

For $x, y, a, b \in \mathbb{R}$,

$$
\begin{aligned}
\max\{\max\{x, a\} + y, b\} &= \max\{\max\{x + y, a + y\}, b\} \\
&= \max\{x + y, a + y, b\}.
\end{aligned}
$$

W. K. Klein Haneveld et al., *Stochastic Programming*, Graduate Texts in Operations Research, https://doi.org/10.1007/978-3-030-29219-5

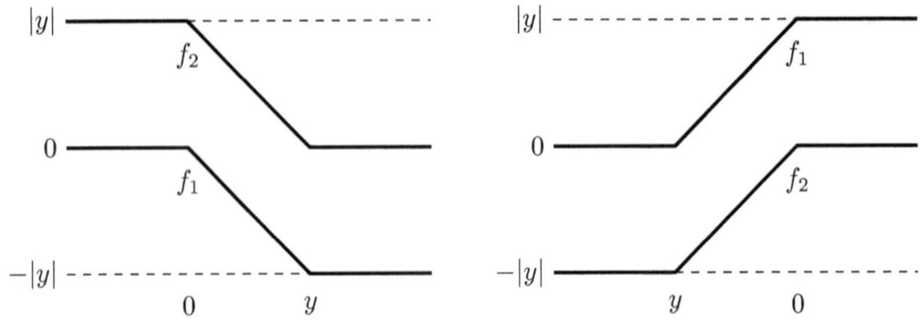

Fig. D.1 The functions $f_1(x, y) = (x - y)^+ - (x)^+$ and $f_2(x, y) = (x - y)^- - (x)^-$ for fixed $y > 0$ (left) and $y < 0$ (right)

Hence, for the special case $a = b = 0$,

$$((x)^+ + y)^+ = \max\{x + y, y, 0\}.$$

Also,

$$\max\{x, a\} + \max\{y, b\} = \max\{x + y, x + b, a + y, a + b\},$$

and for the special case $a = b = 0$,

$$(x)^+ + (y)^+ = \max\{x + y, x, y, 0\}.$$

This formula can be generalized to

$$\sum_{j \in J} (x_j)^+ = \max_{I \subset J} \sum_{i \in I} x_i$$

where $J \subset \mathbb{Z}$ and $x_j \in \mathbb{R}$ for any $j \in J$, and the empty sum is defined as zero. Similarly one obtains

$$\begin{aligned} (x)^+ - |y| \le (x - y)^+ \le (x)^+ + |y| \\ (x)^- - |y| \le (x - y)^- \le (x)^- + |y|, \end{aligned} \tag{D.1}$$

see Fig. D.1.

Note: More often than not, one writes x^+ for $(x)^+$ and x^- for $(x)^-$. This may be confusing, especially in stochastic programming, since + and − are also used as index of a variable or parameter rather than as the operation defined above. In each case one should be aware of the real meaning of the symbols!

Appendix E
Algorithms for Convex Non-linear Optimization

In this appendix we sketch some main ideas behind general algorithms for convex non-linear optimization problems. Some of these ideas will be worked out in more detail in the context of specific algorithms for stochastic linear programming models.

Consider the convex non-linear programming problem

$$\text{(CP)} \qquad \min_{x \in X} G(x)$$

with $G : \mathbb{R}^n \mapsto \mathbb{R}$ a convex (differentiable) function, and the feasible set X,

$$X = \left\{x \in \mathbb{R}^n : g_i(x) \leq 0,\ i = 1, \ldots, l\right\}$$

a *convex closed set*. That is, it is assumed that the functions g_i are quasi-convex and lower semicontinuous (e.g. affine linear or convex). For example, for the model SLPwJCC with joint chance constraints in which only the right-hand sides are random, the feasible set is $X = \{x \in \mathbb{R}^n_+ : Ax = b,\ \alpha - \Pr\{Tx \geq \omega\} \leq 0\}$, i.e., all but the last constraint are linear and $g_l(x) = \alpha - \Pr\{Tx \geq \omega\}$ is quasi-convex if ω is normally distributed.

Without loss of generality we may assume that G is linear. Indeed, by specifying an additional variable $\theta \in \mathbb{R}$ we may obtain the equivalent problem

$$\min_{\substack{x \in X \\ \theta \in \mathbb{R}}} \{\theta : \theta \geq G(x)\}$$

An optimal solution $x^\star$ of CP satisfies $x^\star \in X$ and $G(x^\star) \leq G(x)$ for all $x \in X$.

W. K. Klein Haneveld et al., *Stochastic Programming*, Graduate Texts in Operations Research, https://doi.org/10.1007/978-3-030-29219-5

An algorithm for CP generates a sequence $x^1, x^2, \dots$ for which we hope (or even prove) that $\lim_{k\to\infty} x^k = x^\star$. Only in exceptional cases we will have finite convergence. Many algorithms for CP belong to one of the following two classes:

(1) *Primal methods*: these methods consist of finding an *inner* approximation of $x^\star \in X$, which means that every x^k of the generated sequence is feasible and that 'optimality is improved'. Typical primal methods are e.g. feasible directions method such as Frank and Wolfe's algorithm (see below) and the simplex method for LP.
(2) *Dual methods*: these methods consist of finding an *outer* approximation of $x^\star \in X$, which means that every x^k of the generated sequence is 'optimal' (i.e., dual feasible) and that (primal) 'feasibility is improved'. Typical dual methods are e.g. Veinott's supporting hyperplane method (see below) and the dual simplex method for LP.

Next we present a classical example from either class, indicating some difficulties that arise when applying these methods to SLP problems.

E.1 Frank and Wolfe's Algorithm

In this section we discuss a typical primal method for minimizing a differentiable convex function over a polyhedral set: $\min_{x\in X_0} G(x)$. The method, known as Frank and Wolfe's algorithm, is one of the oldest convex programming algorithms. We deal with it to illustrate basic aspects of primal methods. For actual minimization more advanced methods are available.

The Frank and Wolfe algorithm is applicable to minimization problems with a convex differentiable objective and linear constraints:

$$\min_{x\in X_0} G(x).$$

The objective function G is convex and differentiable, and the feasible set X_0 is polyhedral, e.g. given by non-negativities and $Ax = b$.

Given a feasible starting point x, the algorithm proceeds iteratively as follows:

(a) solve a linear program to find an improving feasible direction, i.e. a direction d in which the objective is decreasing and such that $x + \lambda d$ is feasible for all $\lambda \in [0, \lambda_0]$ for some $\lambda_0 > 0$, and
(b) use linesearch (e.g. by bisection) to find the best solution in the direction found under (a).

At step s of the algorithm, let $x^s \in X_0$ be an approximation of the optimal solution $x^\star$.

(a) Solve an approximating problem, obtained by replacing G by its the linear approximation at x^s

$$G(x^s) + \nabla G(x^s)(x - x^s),$$

which is equivalent to solving the *linear program*

$$\min_{x \in X_0} \nabla G(x^s) \cdot x$$

Let $\bar{x}^s$ be an optimal solution.

(b) Define the search direction $d^s := \bar{x}^s - x^s$. If $d^s = 0$ stop: $x^\star = x^s$. If $d^s \neq 0$, find a solution which minimizes G on the line segment $[x^s, \bar{x}^s]$ by solving

$$\min_{\lambda \in [0,1]} G(x^s + \lambda d^s) =: g_s(\lambda)$$

Note that g_s is a convex function of λ, and that $g_s'(\lambda) = \nabla G(x^s + \lambda d^s) \cdot d^s$ with $g_s'(0) < 0$. This minimization can be performed using bisection: with $\lambda^\star$ denoting an optimal solution (i.e. the optimal step length), if $g_s'(\lambda) > 0$ then $\lambda^\star < \lambda$, if $g_s'(\lambda) < 0$ then $\lambda^\star > \lambda$.

(c) Set $x^{s+1} = x^s + \lambda^\star \bar{x}^s$, $s := s + 1$, and repeat.

Since the search direction d^s is always feasible, the Frank and Wolfe algorithm belongs to the class of *feasible directions methods*. Also, d^s is an improving direction in the sense that the objective G is decreasing in this direction starting from the current solution, so that the method is monotone decreasing. However, d^s is not necessarily a good direction, especially if x^s is still far away from the optimal solution $x^\star$; regularization may be needed.

As is clear from this outline, the method needs many calculations of ∇G: at every step it is needed in part (a) and several times in part (b). Depending on the function G at hand, this may or may not be a problem.

E.2 Veinott's Supporting Hyperplane Method

Veinott's method [50] is a typical dual method for CP. The basic idea is to iteratively

(a) construct a polyhedral outer approximation of the feasible set X,
(b) solve a linear program, and
(c) improve the approximation.

Note that step (b) relies on the assumption that the objective function is linear, which as explained above is no loss of generality.

To explain the method in some detail, it is convenient to write the feasible set X as $X_0 \cap K$, where $X_0 = \{x \in \mathbb{R}^n : Ax = b \ x \geq 0\}$ describes all linear constraints and $K = \{x \in \mathbb{R}^n : g_i(x) \leq 0, \ i = 1, \ldots, l\}$ all other constraints. For simplicity we assume here that the functions g_i are differentiable; this assumption is not necessary.

For the method to work, existence of a $\bar{z}$ such that $g_i(\bar{z}) < 0$ for all non-linear functions g_i describing the set K is needed. Moreover, we must be able to determine such a point.

Exercise E.1 (Assignment C4) How can such a point $\bar{z}$ be determined for models with a joint chance constraint? ◁

At step s of the algorithm we have a polyhedral outer approximation of the set K given by

$$K^s = \left\{x \in \mathbb{R}^n : d_j x \leq e_j, \ j \in J_s\right\},$$

where d_j is a row vector in $\mathbb{R}^n$ and $e_j \in \mathbb{R}$, such that $X_0 \cap K^s \supset X$. At initialization, the set K^0 can be a sufficiently large box (i.e. a set defined by simple lower and upper bounds).

Step s consists of the following:

(a) Solve the current LP problem given as

$$\min_{x \in X_0 \cap K^s} cx$$

Let $\bar{x}^s$ be an optimal solution.

(b) If $\bar{x}^s \in X$ stop: $\bar{x}^s$ is an optimal solution for CP. (Why?)

(c) If $\bar{x}^s \notin X$ select an i with $g_i(\bar{x}^s) > 0$, and construct a *separating hyperplane* $H^s : d_s x = e_s$ such that $d_s \bar{x}^s > e_s$ and $d_s x \leq e_s \ \forall x \in X$.
Set $K^{s+1} := K^s \cap \{d_s x \leq e_s\}$, $J_{s+1} := J_s \cup \{s\}$, $s := s + 1$, and proceed with step $s + 1$.

Thus at each step we need to construct a hyperplane that separates the solution $\bar{x}^s \notin X$ of the approximating problem from X. In Veinott's method this is done as follows:

(i) Find a boundary point y^s of X on the line segment $[\bar{z}, \bar{x}^s]$. Since X is convex, linesearch by e.g. bisection can be used. Let g_i be the constraint in K that is encountered first passing from $\bar{z}$ to $\bar{x}^s$. By construction it holds $g_i(\bar{z}) < 0$, $g_i(\bar{x}^s) > 0$, and $g_i(y^s) = 0$.

(ii) Construct a supporting hyperplane H^s to X at y^s (corresponding to a linearization of g_i at y^s):

$$\underbrace{g_i(y^s)}_{=0} + \underbrace{\nabla g_i(y^s)(x - y^s)}_{\equiv d_s x - e_s} = 0$$

with $d_s = \nabla g_i(y^s) \in \mathbb{R}^n$ and $e_s = \nabla g_i(y^s) \cdot y^s \in \mathbb{R}$. Now $d_s x = e_s$ is a separating hyperplane, since by construction the constraint $d_s x \leq e_s$ is satisfied for all $x \in X$, but not by the current approximate solution $\bar{x}^s$ of $x^\star$.

Given this conceptual outline of the method, some technical problems become apparent:

(1) The method calls for many precise calculations of g_i and ∇g_i, which may be difficult depending on the specification of g_i. For example, if CP is a model with a joint chance constraint with a right-hand side ω that follows a multivariate normal distribution, we need a good method to evaluate $\Pr\{Tx \geq \omega\}$ and $\frac{\partial}{\partial x_j} \Pr\{Tx \geq \omega\}$.
(2) Numerical difficulties in the neighborhood of the optimal solution, caused by the fact that after a while the cutting planes become almost parallel (the problem becomes ill conditioned).
(3) Every current solution y^s is only an approximation of the optimal solution $x^\star$.
(4) At each iteration a constraint is added to the approximating problem, and some constraints may become redundant. To keep the problem size manageable, such redundant constraints should be dropped, but how to detect them?
(5) The performance of the method depends on $\bar{z}$, the interior point of K. How to find a good interior point? Alternatively, how to update $\bar{z}$?

Bibliography

1. K.J. Arrow. *Essays in the theory of risk-bearing*. North-Holland Publishing Co., Amsterdam, 1970.
2. J.F. Benders. Partitioning procedures for solving mixed-variables programming problems. *Numerische Mathematik*, 4:238–252, 1962.
3. C.C. Carøe and J. Tind. L-shaped decomposition of two-stage stochastic programs with integer recourse. *Math. Program.*, 83(3):451–464, 1998.
4. Claus C. Carøe and Rüdiger Schultz. Dual decomposition in stochastic integer programming. *Oper. Res. Lett.*, 24(1–2):37–45, 1999.
5. J. Dupačová. Minimax approach to stochastic linear programming and the moment problem. Selected results. *Z. Angew. Math. Mech.*, 58(7):466–467, 1978.
6. Yu. Ermoliev and R.J-B. Wets. *Numerical Techniques for Stochastic Optimization*. Springer-Verlag, Berlin, 1988.
7. K. Frauendorfer. *Stochastic two-stage programming*, volume 392 of *Lecture Notes in Economics and Mathematical Systems*. Springer-Verlag, Berlin, 1992. Habilitationsschrift, University of Zürich, Zürich, 1992.
8. R.J. Freund. The introduction of risk into a programming model. *Econometrica*, 24:253–263, 1956.
9. R.L. Graham. An efficient algorithm for determining the convex hull of a finite planar set. *Information Processing Letters*, 1:132–133, 1972.
10. C.A. Holloway. *Decision Making under Uncertainty*. Prentice Hall, Englewood Cliffs, 1979.
11. P. Kall. *Stochastic Linear Programming*. Springer-Verlag, Berlin, 1976.
12. P. Kall and J. Mayer. SLP-IOR: An interactive model management system for stochastic linear programs. *Math. Program.*, 75(2):221–240, 1996.
13. P. Kall and D. Stoyan. Solving stochastic programming problems with recourse including error bounds. *Math. Operationsforsch. Statist. Ser. Optim.*, 13(3):431–447, 1982.
14. E.P.C. Kao and M. Queyranne. Budgeting cost of nursing in a hospital. *Management Science*, 31(5):608–621, 1985.
15. S. Katakoa. A stochastic programming model. *Econometrica*, 31:181–196, 1963.
16. W.K. Klein Haneveld. *Duality in stochastic linear and dynamic programming*, volume 274 of *Lecture Notes in Economics and Mathematical Systems*. Springer-Verlag, Berlin, 1986.
17. W.K. Klein Haneveld, L. Stougie, and M.H. van der Vlerk. On the convex hull of the composition of a separable and a linear function. Discussion Paper 9570, CORE, Louvain-la-Neuve, Belgium, 1995.
18. W.K. Klein Haneveld, L. Stougie, and M.H. van der Vlerk. An algorithm for the construction of convex hulls in simple integer recourse programming. *Ann. Oper. Res.*, 64:67–81, 1996.

W. K. Klein Haneveld et al., *Stochastic Programming*, Graduate Texts in Operations Research, https://doi.org/10.1007/978-3-030-29219-5

19. W.K. Klein Haneveld, L. Stougie, and M.H. van der Vlerk. Simple integer recourse models: convexity and convex approximations. *Math. Program.*, 108(2–3):435–473, 2006.
20. W.K. Klein Haneveld and M.H. van der Vlerk. Stochastic integer programming: General models and algorithms. *Ann. Oper. Res.*, 85:39–57, 1999.
21. W.K. Klein Haneveld and M.H. van der Vlerk. Optimizing electricity distribution using two-stage integer recourse models. In S. Uryasev and P.M. Pardalos, editors, *Stochastic Optimization: Algorithms and Applications*, pages 137–154. Kluwer Academic Publishers, 2001.
22. W.K. Klein Haneveld and M.H. van der Vlerk. Integrated chance constraints: reduced forms and an algorithm. *Computational Management Science*, 3(4):245–269, 2006.
23. G. Laporte and F.V. Louveaux. The integer L-shaped method for stochastic integer programs with complete recourse. *Oper. Res. Lett.*, 13:133–142, 1993.
24. F.V. Louveaux and Y. Smeers. Optimal investments for electricity generation: a stochastic model and a test problem. In Yu. Ermoliev and R.J-B. Wets, editors, *Numerical techniques for stochastic optimization*, chapter 24, pages 445–453. Springer-Verlag, Berlin, 1988.
25. R.D. Luce and H. Raiffa. *Games and decisions: introduction and critical survey*. John Wiley & Sons Inc., New York, N. Y., 1957. A study of the Behavioral Models Project, Bureau of Applied Social Research, Columbia University;.
26. H.M. Markowitz. *Portfolio selection: Efficient diversification of investments*. John Wiley & Sons Inc., New York, 1959. Cowles Foundation for Research in Economics at Yale University, Monograph 16.
27. J. Mayer. *Stochastic Linear Programming Algorithms: A Comparison Based on a Model Management System*. Optimization theory and applications; v. 1. Gordon and Breach Science Publishers, OPA Amsterdam, The Netherlands, 1998.
28. R.R. Meyer. On the existence of optimal solutions to integer and mixed-integer programming problems. *Mathematical Programming*, 7:223–235, 1974.
29. G.L. Nemhauser and L.A. Wolsey. *Integer and Combinatorial Optimization*. Wiley, New York, 1988.
30. W. Ogryczak and A. Ruszczyński. Dual stochastic dominance and related mean-risk models. *SIAM Journal on Optimization*, 13(1):60–78, 2002.
31. R.J. Peters, K. Boskma, and H.A.E. Kupper. Stochastic programming in production planning: a case with non-simple recourse. *Statistica Neerlandica*, 31:113–126, 1977.
32. J.W. Pratt, H. Raiffa, and R. Schlaifer. The foundations of decision under uncertainty: An elementary exposition. *J. Amer. Statist. Assoc.*, 59:353–375, 1964.
33. A. Prékopa. *Stochastic Programming*. Kluwer Academic Publishers, Dordrecht, 1995.
34. H. Raiffa. *Decision Analysis: Introductory Lectures on Choices under Uncertainty*. Addison-Wesley, Reading, 1968.
35. R.T. Rockafellar. *Convex Analysis*. Princeton University Press, Princeton, New Jersey, 1970.
36. W. Romeijnders, M.H. van der Vlerk, and W.K. Klein Haneveld. Convex approximations of totally unimodular integer recourse models: A uniform error bound. *SIAM Journal on Optimization*, 25:130–158, 2015.
37. W. Romeijnders, M.H. van der Vlerk, and W.K. Klein Haneveld. Total variation bounds on the expectation of periodic functions with applications to recourse approximations. *Math. Program.*, 157:3–46, 2016b.
38. A. Schrijver. *Theory of Linear and Integer Programming*. Wiley, Chichester, 1986.
39. R. Schultz. Continuity and stability in two-stage stochastic integer programming. In K. Marti, editor, *Lecture Notes in Economics and Mathematical Systems*, volume 379, pages 81–92. Springer-Verlag, Berlin, 1992.
40. R. Schultz. Continuity properties of expectation functions in stochastic integer programming. *Math. Oper. Res.*, 18:578–589, 1993.
41. R. Schultz. On structure and stability in stochastic programs with random technology matrix and complete integer recourse. *Math. Program.*, 70:73–89, 1995.
42. R. Schultz, L. Stougie, and M.H. van der Vlerk. Solving stochastic programs with complete integer recourse: A framework using Gröbner Bases. *Math. Program.*, 83(2):229–252, 1998.

43. C. Schweigman. *OR in development countries*. Khartoum University press, Khartoum, 1985.
44. L. Stougie and M.H. van der Vlerk. Stochastic integer programming. In M. Dell'Amico, F. Maffioli, and S. Martello, editors, *Annotated Bibliographies in Combinatorial Optimization*, chapter 9, pages 127–141. Wiley, 1997.
45. S.P. Uryasev, editor. *Probabilistic constrained optimization*. Kluwer Academic Publishers, Dordrecht, 2000. Methodology and applications.
46. C. van de Panne and W. Popp. Minimum cost cattle feed under probabilistic protein constraint. *Management Science*, 9:405–430, 1963.
47. M.H. van der Vlerk. *Stochastic programming with integer recourse*. PhD thesis, University of Groningen, The Netherlands, 1995.
48. M.H. van der Vlerk. Convex approximations for complete integer recourse models. *Math. Program.*, 99(2):297–310, 2004.
49. R. Van Slyke and R.J-B. Wets. L-shaped linear programs with applications to control and stochastic programming. *SIAM Journal on Applied Mathematics*, 17:638–663, 1969.
50. A.F. Veinott. The supporting hyperplane method for unimodal programming. *Oper. Res.*, 15:147–152, 1967.
51. J. von Neumann and O. Morgenstern. *Theory of Games and Economic Behavior*. Princeton University Press, Princeton NJ, 1947.
52. R.M. Wollmer. Two-stage linear programming under uncertainty with 0–1 first stage variables. *Math. Program.*, 19:279–288, 1980.

Index

W. K. Klein Haneveld et al., *Stochastic Programming*, Graduate Texts in Operations Research, https://doi.org/10.1007/978-3-030-29219-5

MIX
Papier aus verantwortungsvollen Quellen
Paper from responsible sources
FSC® C105338

If you have any concerns about our products,
you can contact us on
ProductSafety@springernature.com

In case Publisher is established outside the EU,
the EU authorized representative is:
Springer Nature Customer Service Center GmbH
Europaplatz 3, 69115 Heidelberg, Germany

Printed by Libri Plureos GmbH
in Hamburg, Germany